# CNC PROGRAMMING AND SIMULATION

A Step-by-Step Guide Practical Approach with CNC Simulator Delcam PowerMILL CAD CAM, and CNC Machining

By

Nishioka Yoshihiro

# TABLE OF CONTENTS

INTRODUCTION TO CNC ................................................................. 5

HISTORY AND TYPES OF CNC MACHINE ...................................... 8

INTRODUCTION TO G CODE AND M CODE ................................. 14

RECAP MODULE KEYNOTES ........................................................ 18

CNC PRO SIMULATOR ................................................................. 21

HOW TO INSTALL CNC SIMULATOR PRO .................................... 23

SOFTWARE INTERFACE ............................................................... 27

SIMULATION CONTROL BUTTON ............................................... 29

SPEED OF SIMULATION .............................................................. 34

PARTS OF SIMULATION .............................................................. 38

CNC EDITOR ............................................................................... 42

PAUSE POINTS ........................................................................... 46

HOW TO MAKE COMMANDS ..................................................... 47

INVENTORY BROWSER ............................................................... 53

MILL AND LATHE WORKPIECE ................................................... 57

HOW TO ADD MATERIAL ........................................................... 61

ZERO POINTS ............................................................................. 63

BASICS OF G CODE PART 1 ......................................................... 65

BASICS OF G CODE PART 2 ......................................................... 68

BASICS OF G CODE PART 3 ......................................................... 72

BASICS OF G CODE PART 4 ......................................................... 75

BASICS OF G CODE PART 5 ......................................................... 78

BASICS OF G CODE PART 6 ......................................................... 80

BASICS OF G CODE PART 7 ......................................................... 86

SIMULATION OF G CODE PART 1 ............................................... 88

| | |
|---|---|
| SIMULATION OF G CODE PART 2 | 91 |
| BASICS OF M CODE PART 1 | 96 |
| BASICS OF M CODE PART 2 | 101 |
| PRACTICAL EXERCISES ON G CODE AND M CODE PART 1 | 106 |
| PRACTICAL EXERCISES ON G CODE AND M CODE PART 2 | 119 |
| PRACTICAL EXERCISES ON G CODE AND M CODE PART 3 | 128 |
| PRACTICAL EXERCISES ON G CODE AND M CODE PART 4 | 139 |
| ADVANCE G CODE AND M CODE | 151 |
| ADVANCE G CODES PART 1 | 157 |
| ADVANCE G CODES PART 2 | 164 |
| ADVANCE G CODES PART 3 | 171 |
| ADVANCE G CODES PART 4 | 177 |
| ADVANCE G CODES PART 5 | 182 |
| ADVANCE M CODE | 188 |
| COMPONENT 1 | 197 |
| COMPONENT 2 | 203 |
| COMPONENT 3 | 211 |
| COMPONENT 4 | 217 |
| CONCUSSION | 226 |
| INFORMATION OF POWERMILL | 228 |
| INTRODUCTION OF POWERMILL | 230 |
| MODEL CREATE AND IMPORT IN POWERMILL | 234 |
| FACING OPERATION-2 | 239 |
| TOOLPATH MOTION OF OPERATION | 243 |
| SET POST PROCESSOR | 246 |
| CREATE NC PROGRAM | 249 |
| POCKET MILLING OPERATION | 252 |

POCKET MILLING OPERATION-DRAFT ANGLE ............... 257

CHAMFER MILLING OPERATION ...................................... 261

CHAMFER MILLING OPERATION-2 ................................ 264

DRILLING OPERATION ...................................................... 266

DIFFERENT DIAMETER DRILLING OPERATION ............... 269

SLOT MILLING OPERATION ............................................. 274

SLOT MILLING OPERATION-2 ......................................... 279

PRACTICAL PROJECT CAVITY PART-1 .............................. 283

PRACTICAL PROJECT CAVITY PART-2 ............................. 288

PRACTICAL PROJECT CORE PART-3 ................................ 292

PRACTICAL PROJECT CORE PART-4 ................................ 295

# INTRODUCTION TO CNC

We will explore the fundamental components of CNC machining and the critical function of the G and M course in the control of CNC machines. We aim to equip you with the necessary knowledge to work with CNC machines efficiently and effectively, as well as to provide you with a comprehensive understanding of CNC machining. We should commence by establishing the foundation for the events that are to come. CNC machining is a fundamental component of contemporary manufacturing, as it enables the automation, efficiency, and precision of a diverse array of industries. Upon completion of this course, you will possess a comprehensive understanding of the fundamental concepts, tools, and protocols that underpin CNC machining. The initial topic of this module will be an overview of CNC machining and its significance in the manufacturing industry. Next, we will address the fundamentals of the CNC, G, and M codes. The subsequent section will address the critical role of G codes in regulating motion and toolpath, as well as the equally critical role of M codes in regulating machine functions and auxiliary operations.

## What will you learn in this Module:

- Overview of CNC machining and its importance in manufacturing.
- Introduction to CNC G and M Codes
- Roles of G codes in controlling motion and toolpath
- Role of M codes in controlling machine functions

Therefore, secure your seatbelts as we commence this expedition to explore the realm of CNC machining and its trajectory. You will be well on your way to achieving proficiency in CNC machining by the conclusion of this module. Before we delve further into the realm of CNC machining, it is imperative that we provide a thorough overview of this revolutionary technology. Initially, we will examine the definition of CNC machining. Here, CNC is an acronym for computer numerical control system, a technology that has significantly altered the manufacturing industry. CNC machining is fundamentally the process of utilizing computer-controlled devices to precisely shape cut all primary materials, including metal, plastic, and wood.

Instructions are meticulously followed by these devices. No one possesses a CNC program that enables them to execute their duties with remarkable precision. When discussing CNC machining, it is common to have the query of why these devices are so significant. Therefore, the initial aspect of CNC machining is its importance: precision. CNC machines provide an unparalleled level of precision, enabling the production of intricate and complex parts with minimal error. The second component is automation. It results in a reduction in the necessity for manual labor, which in turn enhances the consistency and efficacy of production. The third characteristic is adaptability. Therefore, CNC machining is implemented in a diverse array of sectors, including aerospace, automotive, medical device manufacturing, and beyond. Therefore, this is the fundamental information concerning the significance of these CNC machines. Additionally, in relation to the concise historical perspective and the CNC machines that are employed in the manufacturing process.

# HISTORY AND TYPES OF CNC MACHINE

We will engage in a concise historical examination of CNC machining and the various CNC devices that are employed in the manufacturing process. In order to comprehend the assessment of CNC machining, it is necessary to write a concise summary. In general, the concept of numerical control can be traced back to the 1940s, when engineers began experimenting with punch card systems to regulate machine tools. The first genuine CNC machines emerged in the new era of manufacturing in the late 1950s and early 1960s. The integration of cutting-edge technologies such as artificial intelligence and the Internet of Things has resulted in the ongoing evolution of CNC machining. In conclusion, CNC machining is the foundation of contemporary manufacturing, facilitating automation and precision on a scale that was previously unimaginable. Subsequently, we will investigate the various varieties of CNC machines and investigate the advantages of CNC manufacturing in comparison to conventional manual methods.

# CNC Machines in Manufacturing

Therefore, please remain with us as we continue our exploration of the CNC industry. CNC technology is exceedingly adaptable and is implemented in a variety of devices, each of which is customized to execute a particular function. The following are some of the most frequently encountered categories of CNC machines. The initial category is CNC milling devices. Typically, these devices are employed to cut and shape materials, such as metals or plastics. In order to eliminate material from the workpiece, they implement rotating cutting instruments. The second category is CNC milling devices.

Types of CNC Machines:

CNC Milling Machine

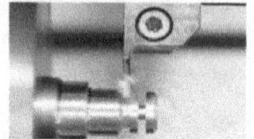
CNC Turning Machines

CNC Drilling Machines

CNC Plasma-Cutting Machine

CNC Routers

Cylindrical components are manufactured using turning machinery. The workpiece is rotated while the cutting instrument is used to mold it. The CNC drilling devices are the third category. These devices are employed to drill holes into materials with precision and consistency, as their name implies. The fourth item is the CNC laser and plasma cutters. Ideal for intricate shapes and fine details, these devices employ high-powered lasers or plasma beams to cut through materials. The CNC routers are the final item. These devices are employed in the woodworking industry and are capable of carving intricate patterns and designs into wood. Therefore, any of the machines that are enumerated here can be used by anyone, provided that they meet the necessary criteria. At present, you will observe the amalgamation of numerous machines that have integrated the capabilities of various machines. Subsequently, it is imperative to comprehend the rationale behind the preference for CNC manufacturing over conventional manual methods. In order to accomplish this, it is necessary to comprehend the benefits of CNC machining. Precision is the initial benefit. Therefore, CNC machines are capable of achieving precision levels that are virtually impossible to replicate through manual machining. This precision guarantees the production of elements that are exceedingly precise. The second advantage is the ability to replicate the process. CNC devices are capable of reproducing these operations with consistent precision. They mitigate production variability. Next, efficiency is a critical advantage. Automation results in a decrease in labor costs and an increase in production efficacy.

Advantages of CNC Machining

1. Precision
2. Repeatability
3. Efficiency
4. Complex Geometry
5. Material Versatility

CNC machines can operate continuously for 24 hours a day with minimal supervision. The subsequent topic is complex geometry. CNC machining enables the production of intricate and complex shapes and designs that are frequently unattainable by manual machines. Material versatility is the final aspect. Therefore, CNC machines are capable of operating with a diverse array of materials, such as metals, plastics, ceramics, and composites. Additionally, the machining of certain rigid materials, including ceramics and composites, is exceedingly challenging. Through the utilization of manual devices. Therefore, these CNC machines are exceedingly advantageous for the machining of these materials in this location. Subsequently, we will examine the implementations of these CNC machines in various aerospace industries. CNC devices are employed to fabricate essential components for spacecraft and aircraft. The automotive sector places a high value on precision and reliability. The intricate designs on the dashboards, as well as engine components and transmission elements, are all manufactured using CNC machining. CNC machines are capable of fabricating intricate, high-precision components, including surgical instruments and implants, in the medical device manufacturing industry. This category of surgical instruments and implants necessitates a high degree of precision and accuracy.

### Applications Across Industries
1. Aerospace industry
2. Automotive sector
3. Medical device manufacturing
4. Electronics industry
5. Artistic and Architectural field

In contrast to manual machines, CNC machines are exceedingly advantageous. CNC machining is essential for the production of custom enclosures and printed circuit boards in the electronics industry. Even in the realms of architecture and art. Intricate architectural details and sculptures are produced by CNC routers. In conclusion, CNC devices are the backbone of contemporary manufacturing. Their versatility, precision, and reproducibility have rendered them indispensable in a diverse array of industries, thereby influencing the future of manufacturing and fostering innovation. In the subsequent sections, we will delve into the realm of CNC programming and investigate the language of CNC through G and M interfaces. Therefore, please remain attentive as we reveal the intricacies of CNC programming.

# INTRODUCTION TO G CODE AND M CODE

As we have developed a more profound understanding of CNC machines, it is now time to explore the fundamental language that powers these machines. Additionally, these are the CNC, G, and M codes. The fundamental components of CNC programming are CNC, J, and M codes. Then, consider them as the alphabet and grammar options that communicate instructions to CNC machines in a standardized language. Machining. Let us examine the definition of CNC codes. Therefore, the CNC G-code is a representation of the geometry codes. It also specifies the type of motion and toolpath that the machine should follow. Initially, we will investigate the function of g codes. If G00 is entered into this machine, it is intended for swift positioning. This code is employed for rapid positioning without machining, as it directs the machine to move swiftly to a designated location. Similarly, if we denote G01 as "linear interpolation," it signifies that.

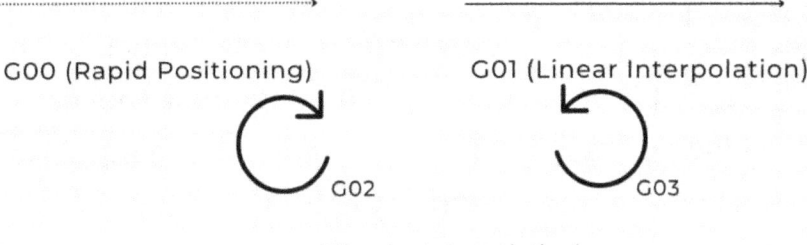

This code directs the machine to proceed in a straight line between the two points, enabling the creation of precise contours and straight incisions. Consider the subsequent illustration. That is, the circular interpolation is indicated by the values of G02 and g zero, which are three and eight. Therefore, this code permits the machine to operate in a circular or arc motion. G02 is advantageous for clockwise motion, while G03 is advantageous for clockwise motion. In our subsequent lectures, we will become acquainted with these G codes in greater detail. Once we have a comprehensive understanding of D and g codes. Subsequently, the accuracy of the machined part is contingent upon the correct selection and sequence of G codes in order to achieve the desired tool. The function of G code in defining the toolpath is one of its most critical components. The cutting tool's precise path as it traverses the workpiece is referred to as the toolpath. This component is essential for the precise and efficient contouring of the workpiece. The tool is guided along the predetermined path by G-code, which guarantees that the material is cut or saved precisely as intended by the machine. It is imperative to comprehend the tool path planning process in order to optimize machining operations, reduce material waste, and achieve high-quality results. We will now discuss the d cnc m codes. Machine Lineas Codes (CNC m code) are responsible for the control of a variety of machines, functions, and auxiliary operations. Consider specific instances, such as M0 three. Therefore, this code is beneficial for initiating the spindle. The spindle rotation is initiated by this code, which enables the tool to cut or save the workpiece. The M05 is the subsequent core, and it is advantageous to halt the spindle in order to prevent its rotation. Capturing the chopping action. If we enter M08, this code will be beneficial for initiating the refrigerant supply. Therefore, M08 activates the coolant system, which is crucial for the dissipation of heat and the extension of the tool's lifespan. Therefore, the M code is of paramount importance in the supervision of machine functions that are indispensable to the machining process. Next.

# M Codes = Miscellaneous codes

M03 (Spindle Start)   M05 (Spindle Stop)   M08 (Coolant On)

In CNC machining, safety is of the utmost importance, and M codes are instrumental in guaranteeing the safe operation of the machine. For example, the M05 command is beneficial for stopping the spindle, which can be crucial in emergency situations or when tool adjustments are necessary. M codes are essential for the coordination of various machine functions throughout the CNC program, in addition to their use in individual operations. The machine operates safely and efficiently as a result of their proper use, which prevents errors and accidents. In conclusion, M codes are the command signals that regulate the critical machine functions, thereby guaranteeing the safe and efficient operation of CNC manufacturing. It is essential to comprehend the appropriate use of M codes in order to achieve the intended machining results and to ensure a safe work environment.

# RECAP MODULE KEYNOTES

As we conclude this module, it is important to review the primary insights and conclusions that you have acquired thus far. These summaries will function as a concise reminder of the fundamental concepts that have been addressed in this module. We commence by underscoring the importance of CNC machining in contemporary manufacturing. CNC technology is a fundamental component of a variety of industries, including aerospace and medical device fabrication, due to its precision, efficiency, and adaptability.

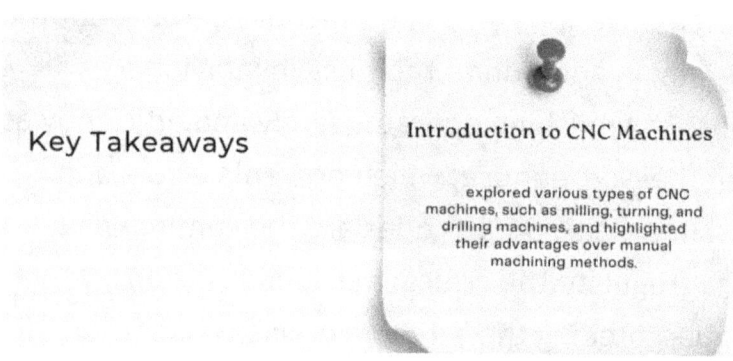

Key Takeaways

Introduction to CNC Machines

explored various types of CNC machines, such as milling, turning, and drilling machines, and highlighted their advantages over manual machining methods.

We investigated a variety of CNC machines, including milling, turning, and piercing machines, and emphasized their advantages over manual manufacturing methods. You have been introduced to the fundamental language of CNC programming, which includes G codes for motion and toolpath control and M codes for machine function administration. The primary significance of G codes was thoroughly investigated, with an emphasis on their ability to regulate tool movement and influence the toolpath. The precision with which the instrument is guided is demonstrated by examples such as the zero zero, G01, G02, and G03. Therefore, the G01 command will be beneficial if you desire a direct movement, as per the requirement. Additionally, the G02 and G03 commands will prove advantageous for circular movement. In the same vein, we investigated M codes and their role in regulating machine functions and guaranteeing safety during the machining process. Their significance in machine operation was illustrated by examples such as M0 three, M05, and M08. Now, as we are aware, G codes are essential for the definition of tool parts, as they direct the cutting tool's precise path across the substrate. These controls are essential for attaining high-quality results, efficiency, and accuracy, as the computer executes the entire operation based on the commands it receives. Determination of the appropriate instrument path. It is crucial to employ these G codes. To manage a variety of machine functions and auxiliary operations, M codes are once again indispensable. This code guarantees the safe and efficient operation of the machine. After completing this module, I trust that you have developed a solid understanding of CNC machining and the fundamental protocols that operate CNG machines. We have only recently addressed the significance of g code in m code in d CNC machines.

# Introduction to CNC Code Simulation

**Why CNC Code Simulation Matters:**

- Cost Savings
- Efficiency
- Safety
- Skill Enhancement

This knowledge will be further developed as we advance in this series. I will delve further into the CNC programming and examine more advanced concepts. In the subsequent module, we will investigate the fundamentals of CNC programming, integrating the theory into practical applications through exercises and demonstrations. So prepare to dig deeper into the realm of CNC and roll up your sleeves.

# CNC PRO SIMULATOR

In the CNC Code Simulation module, we will explore a variety of CNC programming skills that can be improved through hands-on virtual practice. Initially, it is imperative that we comprehend the fundamental objectives of this module. Initially, I will provide you with an overview of the CNC simulation software. You will acquire an understanding of its nature, the reasons for its importance in CNC programming, and its capabilities. The manner in which you approach CNC machining. Subsequently, I will provide you with instructions on how to execute GNN core programs in a virtual CNC environment. This practical experience will allow you to observe the CNC execution, observe tool movements, and witness your code come to life. As we are aware, none of these software applications are bug-free. There is no CNC programming voyage that is complete without addressing errors and collisions. I will provide you with the necessary knowledge and resources to identify common errors that may arise during the machining process and to effectively resolve them. Initially, you may be curious as to why CNC code simulation is such a critical skill. In order to comprehend this, let us first examine some of the critical aspects that are significant. The simulation of the CNC core.

# CNC Simulator Pro

1. Virtual Machining Environment
2. G and M Code Support
3. Realistic Simulation
4. Error Detection and Collision Avoidance
5. Code Verification
6. Toolpath Optimization
7. Multi-Axis Simulation
8. Post-Processor Compatibility
9. Learning and Training
10. Cost Savings
11. Compatibility

First and foremost, there is the cost savings. By detecting errors and collisions in a virtual environment, time and resources are conserved, and costly errors on real machines are averted. As we are aware, the cost of numerous materials is exorbitant, and the machining of lifeless materials necessitates the use of weighty tools. Any errors made during the task will result in increased costs and time loss. Therefore, it is more advantageous to simulate this process. The efficiency simulation is the next critical aspect, as it enables you to optimize your programs for optimal efficiency, thereby reducing material waste and machining time. Safety is the subsequent critical aspect. Therefore, it is possible to establish a secure work environment for both the operator and the devices by comprehending collision detection and avoidance. The final and most critical aspect is the development of skills. Therefore, your CNC programming abilities are elevated to a professional level by mastering code analysis and error resolution. In both academia and industry, comprehension of CNC code simulation will prove advantageous. Additionally, we will now explore the captivating realm of CNC code simulation. The CNC simulation will be conducted using the CNC Simulator Pro software in our subsequent sessions. CNC Simulator Pro is a user-friendly and exhaustive software application that is specifically designed for the simulation and testing of CNC programs. Consequently, no CNC programming journey is complete without addressing errors and collisions. It is a valuable instrument for CNC programmers, machinists, and anyone involved in CNC manufacturing due to its numerous features and benefits.

# HOW TO INSTALL CNC SIMULATOR PRO

Let us explore the captivating realm of CNC core simulation. The CNC Simulator Pro software will be employed in our subsequent sessions to simulate this scenario. CNC Simulator Pro is a user-friendly software application that is adept at simulating and evaluating CNC programs in a virtual environment. It is a comprehensive SIM. It is a valuable instrument for CNC programmers, machinists, and anyone involved in CNC manufacturing due to its numerous features and benefits. Therefore, we should initially examine some of the features and advantages of these software programs. The initial one is the virtual machining environment. CNC Simulator Pro offers a virtual machine shop that enables users to generate, modify, and simulate CNC programs without the necessity of a tangible CNC machine. A 3D representation of a CNC machine, workpiece, and tool is available for users to interact with, providing them with a realistic understanding of the machining purposes. The second aspect is the support for G and M codes.

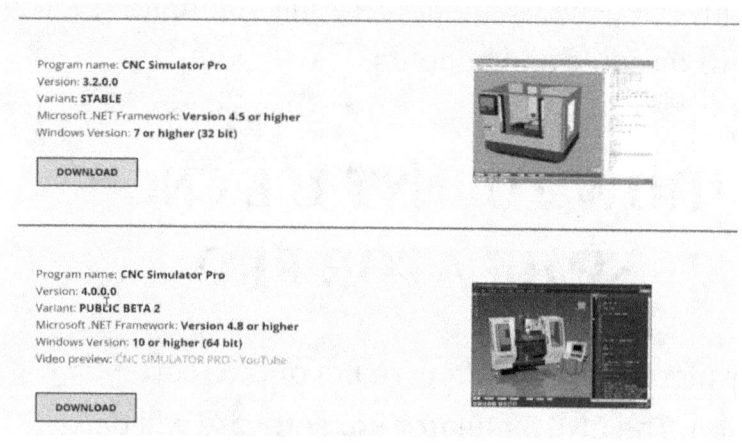

The software supports both G and M codes, enabling users to simulate and test programs that are written in these languages. The device is compatible with a broad variety of CNC machine models and controllers, as it recognizes a wide range of standard G and M codes. A highly accurate and realistic simulation of CNC machining operations is provided by CNC simulator Pro, allowing users to observe the toolpath tool movements and material removal in real time. It replicates a variety of manufacturing operations, such as milling, turning, and binning. One of the most notable features is its capacity to identify errors and prospective collisions during the simulation. Toolpath errors, syntax errors in the code, and collisions between the two workpiece and machined components are among the issues that users can identify. Before conducting the simulation, the software validates the CNC code for syntax errors and correctness. It assists a user in identifying errors at the outset of the programming process. The CNC Simulator Pro also enables the user to optimize tool parts for quality and efficiency. It offers the ability to modify feed rates, spindle speeds, and tool adjustments in order to accomplish the desired outcome. For CNC machines that have multiple axes of movement, such as those with four or five axes. The software is capable of supporting multi-axis simulation and programming. The software frequently includes pre-installed POS processors or enables the user to create custom post processors to convert simulated programs into machine-specific code. CNC Simulator Pro is an exceptional educational resource. It is frequently implemented in CNC training programs to instruct students on the fundamentals of CNC programming and machining. It provides a risk-free environment for learners to experiment and practice with CNC programming concepts. The CNC simulator Pro enables users to avoid the costly errors and material wastage that can result from the execution of unverified CNC programs on real machines. This software is frequently compatible with a variety of CNC machine brands and controllers, which renders it adaptable and versatile in various configurations. In conclusion, CNC Simulator Pro is a potent instrument that serves as a conduit between the theoretical realm of CNC programming and the practical application of machining. It provides a secure and efficient environment for users to develop, test, and optimize CNC programs, thereby improving their productivity and skills in CNC machining operations. The software interface will appear on your screen after the software is launched. Therefore, it is imperative to verify that all components are configured appropriately prior to commencing CNC simulation with these CNC Simulator Pro. Therefore, in the subsequent segment, we will guide you through the process of configuring the software to optimize the effectiveness of our hands-on CNC core simulation session.

# SOFTWARE INTERFACE

I trust that you have successfully downloaded and installed the CNC Simulator Pro. To initiate the software, simply double-click the CNC Simulator Pro icon. The software interface should now appear. And this will be very comparable, as you can see on your screen. Initially, it is necessary to become acquainted with the software interface. The 3D machine model is among the numerous critical components that are typically present. This 3D machine model is the visual representation of the CNC machine with which you will be working, as shown on the screen. The menus and toolbars are also present. This includes a variety of functions and options for program management, simulation, control, and more. On the right-hand side, you will locate the program Editor, which is the location where you can input, modify, or import CNC code programs. The CNC simulation controls are located at the bottom. Additionally, these switches enable you to regulate the simulation by controlling its start, press, and stop.

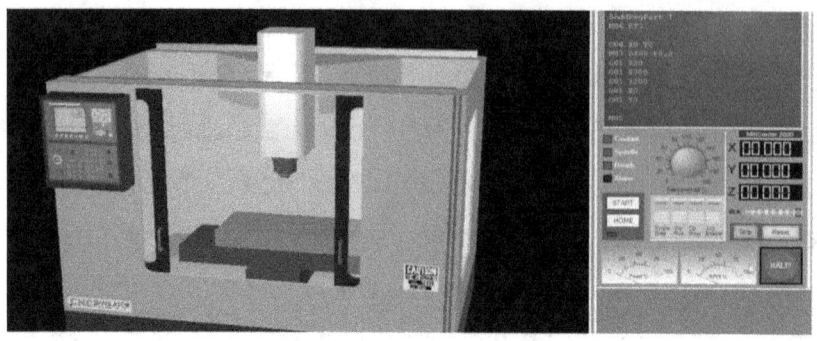

The controller option, which is the status indicator, is located on the right-hand side, in close proximity to the editor option. It is essential to monitor these indicators, as they display critical information about the simulation, including the tool status, input rates, and spindle speed. Similarly, these simulation parameters exhibit the same characteristics. Here, you can observe the tool's status, feed rates, spindle velocities, and movement in accordance with the selected coordinates. Let us first comprehend the process of controlling the 3D view and this 3D view. I recommend that you utilize the mouse to zoom in and out on this machine by rolling the mouse wheel. Therefore, by employing the zoom in and zoom out features, it is possible to obtain a more comprehensive perspective on the component that is being developed. In this 3D interface, you may also right-click on the cursor and drag it to pan the image. Therefore, employing these right-click mouse actions. The machine can be readily moved to meet your needs, and the view can be rotated by left-clicking and dragging. Therefore, by utilizing these left-click functions, you can rotate the component to your liking in order to better view it. Therefore, the utilization of a mouse will be highly beneficial in obtaining a more comprehensive view of the component from various angles and perspectives.

# SIMULATION CONTROL BUTTON

Therefore, as we previously discussed, the simulation control icons are located at the bottom. Therefore, we will now delve into the intricacies of each icon that is present in this section, working our way from left to right. Therefore, when you position your cursor over the initial icon that signifies the commencement of the simulation. This is the play button, which is utilized to initiate the simulation and to recommence it after a delay. The delay icon is the subsequent icon that is accessible. Additionally, this allows you to resume the simulation. The third icon available is the option to halt or reset the simulation. Additionally, the simulation may be terminated by employing this third icon. The fourth icon denotes the rapid forward simulation, which is utilized to simulate at a high speed and is primarily restricted by the hardware of your computer. The subsequent icon is the "step" indicator, which is utilized to progress through the CNC program in a step-by-step manner. These functions are also accessible from the simulator menu, which is located at the top of the screen on this simulate option.

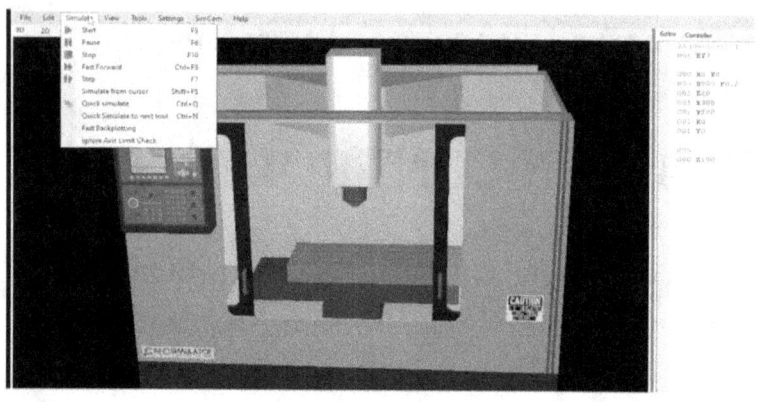

Therefore, you will also discover supplementary functions, including the step option, rapid forward, halt, and stop, when you select this option. These are comparable to those that are accessible at the bottom. In addition to these, there are a few supplementary functions. The initial function is the "simulate from cursor" function. It is employed to initiate the simulation from any location within the CNC program. The Quick Simulate is the subsequent option, which is employed to simulate the CNC code at the highest possible speed. The CNC code will remain concealed in the editor. The subsequent alternative is expedited. Execute a rapid simulation of the subsequent tool to facilitate the subsequent tool change. The next option is rapid back plotting, which is particularly advantageous when working with extremely large CNC programs, as it generates only the tool toolpath. The final alternative is to disregard the axis limit check. Therefore, it is a unique function that causes the simulator to disregard the limits of the axes. Please be advised that the keyboard auxiliary keys, such as F5, may be employed to initiate the process. In the same manner, you will be presented with a list of various keyboard commands that are available for use, including Option F6 for pause, F10 for top, and Control + F5 for rapid forward. Currently, a sample code is available on the right-hand side. To initiate the simulation, you may either select the "start" option from the "simulate" menu or navigate to the bottom and select the "play" icon. Therefore, we should choose it. Upon selecting it, the machine will commence executing the code and execute the operation in accordance with the written code. We will once more select this play icon. And now, presume that we wish to halt the code and select the supervisor in the interim. Upon selecting it, a message will appear with a red highlight, instructing you to post and play in order to continue the simulation. Therefore, in order to resume this operation, you must select the "play" icon. Therefore, I will once more select it. Now, you will observe that the remaining operation has been finalized. In the same manner, you may utilize any of the icons that are available on this website in accordance with your needs. I will once more select the "play" icon to halt the simulation. Suppose that I wish to halt or restart the simulation after it has been paused. Subsequently, select this third icon. Consequently, upon selecting it, the highlighted stages of operation designated as Z will be eliminated from the right-hand side. Therefore, you will no longer have the option to continue the operation.

Therefore, it is necessary to proceed by tapping on the perform simulation icon once more. Finally, you have the option to fast-forward a simulation. We should choose it. Therefore, upon selecting it, you will observe the speed at which the operation is completed in accordance with the selection. Select the single-step simulation option. Therefore, I will select it only once. Therefore, the initial line is executed and the participants are incorporated into the CNC machine on the right-hand side. We will once more select the subsequent simulation. Therefore, you will notice that the code is executed line by line on the right-hand side. Therefore, we should select the "step-by-step simulation" option and execute the entire operation. The subsequent operation is currently being executed. In the same manner, upon clicking on this step-by-step, you will observe that a distinct operation is executed, and you will locate the same post by clicking on the continuation button. If you wish to rerun the entire simulation. Lastly, you must select the "Start Simulation" icon. Therefore, I shall opt for it. The operation is ultimately concluded at this point. Presently, in order to obtain a more comprehensive examination of the machine, the component is examined. Let us utilize the mouse to zoom in. Here, you will observe that the rectangular section of this part is dematerialized in accordance with the selected tool and the dimension provided. The tool movement can be readily observed by following the tool path, which is indicated by the green color.

# SPEED OF SIMULATION

The CNC Simulator Pro allows for the simulation of a variety of velocities. Your hardware determines the utmost simulation pace. Right here. Please be advised that the speed of the simulation is not indicative of the actual machining time. For example, a component that may require an hour to produce in a physical machine may be completed in a mere 10 seconds in this simulator, contingent upon the hardware and speed setting. Therefore, in order to regulate the simulation pace, you may navigate to the bottom of the simulation control icons, where you will locate the button with the spectacles designating the view panel. Therefore, choose it. Upon selecting it, you will observe that a variety of options are available on the left side of this interface. You have the ability to specify which components you wish to observe during and after the simulation. The workpiece design can be selected on the right-hand side. The materials you have specified in the inventory browser are the source of the design, which we will elaborate on in our subsequent sessions. The simulation speed controller is located on the right side of the panel, and it can be utilized to regulate the simulation's pace. Therefore, the slider will be displayed in a variety of hues. Therefore, the toggle is presently functioning as a top. That is, it is chosen.

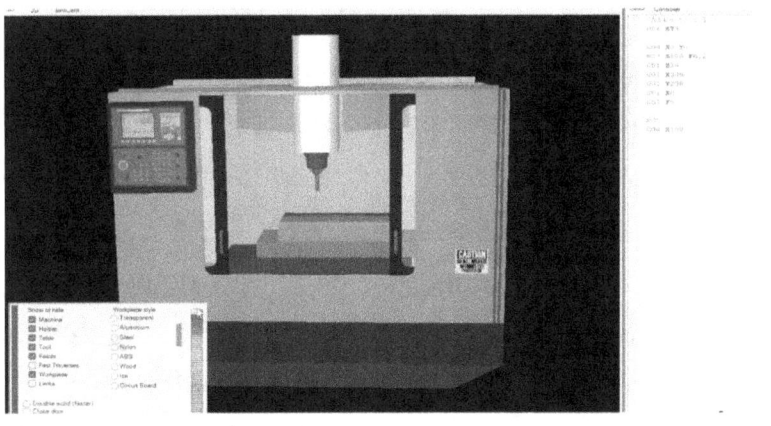

The simulation is operating at full pace. For the most rapid simulation speed, it is recommended that you adjust this slider to the maximum setting of 200%. First, we will execute the simulation operation using the sample code and evaluate the simulation performance. Therefore, I will select the "play" icon and presently monitor the simulation pace in accordance with the data that has been supplied. We will now return to this view panel. Now, select these glasses and drag the slider to the bottom. Therefore, if you position this slider in the center of these vertical lines, it will indicate a 50% simulation speed. Therefore, we should reduce it. I will now click on the play icon once more. These items demonstrate that the residence has been terminated. The pace of the simulation is. Therefore, by maintaining this slider position, you can regulate the simulation pace to your liking. The reduced speed can be employed to observe the simulation operation with great precision. In an effort to enhance our perspective, we shall employ the magnify feature. And here, you can observe that this is rotating and eliminating the material from the workspace. Once more, we should alter our course. Therefore, it is evident that the basic material is eliminated in accordance with the code that has been supplied. I trust that you have gained an understanding of the methods by which the simulation pace can be controlled in this CNC simulator Pro.

Therefore, I will halt the simulation at this point. Once more, expand your view to obtain a more comprehensive perspective of this device. Once more, select this view panel. On the left-hand side, you will observe that various options are selected. Therefore, if you wish to observe the machine, you may do so. Alternatively, if you wish to eliminate this machine, select the "Remove" option. This is where you will observe that no CNC machine is concealed. In the same manner, you can uncheck the holder to conceal it, as well as the table and tool. Therefore. However, it is recommended that you obtain a view that is more favorable. You have the option to select these items, including machine collider, table, and tool. Similarly, it is recommended that you examine this input option in order to thoroughly verify the path. Additionally, the path that this utility follows will no longer be highlighted in the machine if it is unchecked. Therefore, it is recommended that you examine these disciplines in order to obtain a more comprehensive understanding of the path. Additionally, there is an alternative. In the same vein, you may verify this compartment of the workpiece to obtain a more comprehensive view of it. The workpiece design may be specified on the right-hand side. Additionally, if you desire a transparent workpiece, you may select it. In the same way, you will be presented with a variety of options, including aluminum and steel, which can be chosen to create a unique viewing experience in your workspace. Therefore, I will exclusively choose this aluminum at this time. Therefore, the following is a comprehensive inventory of the various workpiece styles that are currently available.

# PARTS OF SIMULATION

We will now explore the subsequent option that is situated on the right-hand side of the current. Additionally, this is the Knife Save icon. The workstation cutting interface can be accessed by clicking on the knife icon-adorned button. You will observe that the left and right fronts are equipped with a variety of adjustments. Make and the upper portion. These sliders can be employed to cut away from any angle in order to observe. These will enhance the appearance of its interior. We will once more employ the Pan and Zoom tool. Use the left-click to zoom in once more. Relocate it to obtain a more favorable perspective. We will now utilize the sliders that are located on the left-hand side. So, we will move this slider in conjunction with the movement of this slider. The part automatically acquires this x from the left side, as is evident. In the same vein, we should endeavor to navigate to the right. Consequently, the modifications are visible in this location. The same principle applies if you desire a sectional appearance from the front. Afterward, you may utilize the third variable that is accessible. Therefore, you will once more behold the distinction in the sectional view.

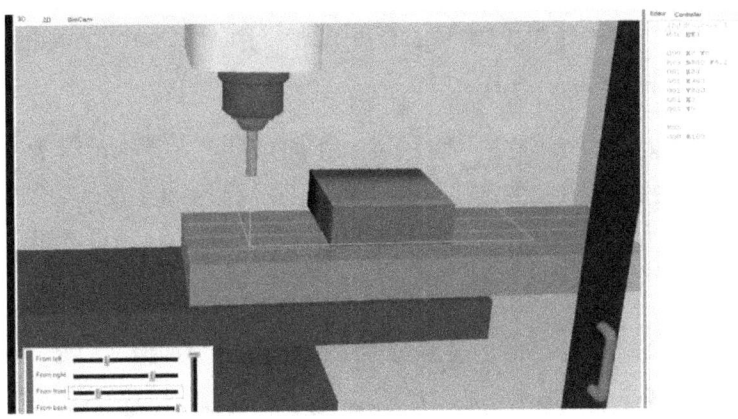

In the same manner, it is possible to obtain a view from the rear. Additionally, in accordance with your specifications. The specific section of the workpiece is visible as the slider moves. Similarly, we can attempt to relocate the toggle that is currently accessible on the upper side. Consequently, the height of these workpieces can be observed as the lever is moved again. Therefore, this knife icon is beneficial for obtaining a more comprehensive understanding of the various cut sections. While possessing intricate geometry. I would like to provide some essential nods now, as you are cognizant of this solid part simulation and this PID. The simulation component necessitates a significant amount of memory and computational capacity. The efficacy will be contingent upon the system's hardware and the simulation's speed. Utilize the rapid forward icon to accelerate the process. Additionally, we have identified the location of the rapid forward option. In the same vein, to ensure the quickest simulation, select the "Disable Solid" option in the view interface. Therefore, navigate to this view panel and select this option from there. To expedite the simulation, disable solid. This is frequently the case, as it is contingent upon the code that is accessible for the machining process. The robust simulation may not be feasible due to the inadequacy of your computer or graphics card. In that event, it may be necessary to disable solid simulation altogether by selecting the Disable Solid checkbox in the view interface. You will continue to have the ability to simulate machine movements and tool parts. A solid buffer resolution selection is also available in the program pane. In this settings editor. Therefore, we should initially navigate to these parameters in the menu. Choose it. Navigate to the option preferences. For the first time, the program is presented here. Therefore, this tab allows you to implement specific modifications in accordance with your specifications. The initial factor is the hue of the background. The background color is black by default in this instance. However, you are free to choose any available option from the color program in accordance with your needs.

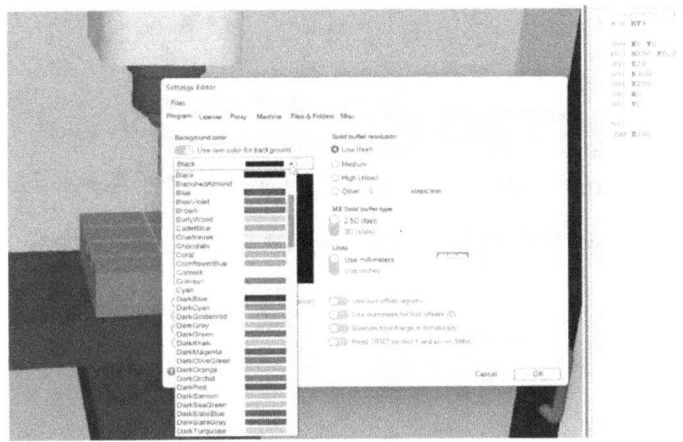

Similarly, the solid buffer resolution is located on the right-hand side, and it can be configured to be low, medium, high, or any other value. Set the resolution to low Y servers if the simulation operates slowly. If you possess a capable computer, you may opt for medium or high settings to simulate intricate solid components. You have the option to input your own buffer resolution value by selecting the last option and the least that is the other. However, I suggest that you select the low, medium, or high option according to your needs. Mainly solid buffer is an option that is located at the bottom of this solid buffer resolution. Then, you may configure it to either 2.5D or 3D. Similarly, the default setting for the machining purpose in the code is three millimeters or 2D. If you wish to swap to a different dimension, simply click on the icon that corresponds to the desired dimension. Additionally, to transition to three millimeters, select the identical icon once more. Therefore, this is where you can establish the element for your program. At present, we are not interested in making any modifications. Therefore, select "okay." As you are aware, the CMC Simulator Pro offers a variety of features that are crucial for our programming.

## CNC EDITOR

Write and modify your C and C code in the CMC editor, which is situated on the right-hand side. It includes functions that are pre-installed to aid in the programming process. Recognize that the n c codes will be highlighted in a pop-up list, which will provide you with valuable information to assist you in your writing. For instance, if I have already composed the code in this location, you will observe that the M codes are highlighted in a red color. The window will automatically appear once the cursor is moved to this location. In the same vein, the inventory of various hues will be presented here. Regarding this, we will engage in a step-by-step discussion. Initially, we will address the topic of code coloring. Therefore, the DCN C editor will automatically color items that it recognizes. For instance, all G codes that are recognized will be shown in cyan, as illustrated here. The color of all M codes will be dark red. The spatial Simulator Pro commands are represented by a dollar symbol and are colored gray when you begin to use the various essential commands. Again, if you wish to include a comment, it will be highlighted in azure. Suppose that I wish to include the command in the second line and refer to it as the first C and C program. Therefore, the command will be referenced between these round brackets and between these round brackets. Assume that I have written the b and c programs first. The command is written between the brackets, as you can see. Additionally, it will manifest as emerald green. In addition, it is important to note that the color will not be displayed when the program is extremely large, such as those used for 3D printing at the time, as it will cause the editor to slow down. However, in the majority of our cases, you will observe these distinct color combinations. Subsequently, we will investigate the extent to which the pop-up will facilitate our programming. Let us now consider the possibility of utilizing this specific command after this line. The m code name will be referenced in an automatically generated pop-up window when I enter M. Subsequently, it is another viewpoint. Therefore, these pop-up windows will be extremely beneficial in enabling the selection of the appropriate code in accordance with the specifications. Therefore, the pop-ups will be of assistance to you during the coding process. Therefore, if you are aware of the content to be written, you may disregard the pop-up list and proceed with the composing process. However, these pop-ups will be extremely beneficial if you are unfamiliar with which command is most beneficial for a particular form of task, which will be particularly helpful for us as beginners. Simulator Pro will display an assistance tooltip window that provides a detailed explanation of any G or M code that is on the list and is recognized by the agency. If you wish to navigate up and down this list, you may utilize the arrow keys on your keyboard. As you do so, a tooltip window will automatically appear, providing a detailed explanation of the code. In the event that we are unaware of the M03 code, it will be evident that these codes are beneficial for initiating the anticlockwise rotation. The code for these codes is M03. As an alternative, you may select a code from the list by tapping on it with your mouse. For example, if I wish to employ the M08 code, I would simply transfer the cursor to this location and select on the M08. Again, the tooltip window will appear to provide a detailed explanation of the code. The code will be inserted into the editor at the cursor location if you double-click on it, press enter or tab on the keyboard, or select it.

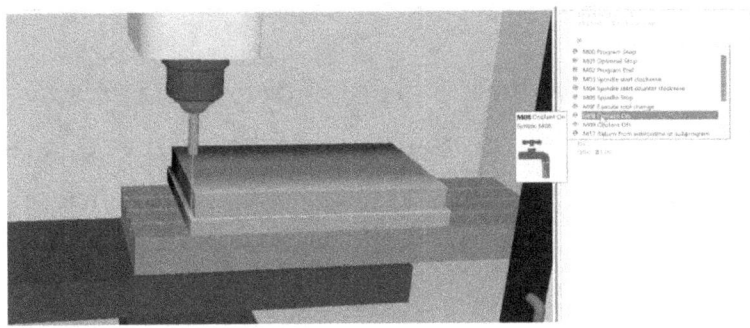

Therefore, we will select the enter key, and you will observe that the M08 code has been incorporated into the code editor. The overall concept of the intelligent editor is to facilitate the user's ability to promptly locate the codes they wish to employ. Another method of aiding the programmer is to display the tooltips when the cursor is hovered over the recognized code. Therefore, I strongly advise that novices utilize this feature when programming a CNC machine.

# PAUSE POINTS

The post markers are these tiny green dots in the editor margin that are utilized to create a post during the simulation. To add or remove a post point, simply click on the margin in the front of the block where you wish to post. In the event that I wish to write a post at these g0 1X3 hundred, I will select the margin on the left side of the blog and locate the green post points to resume the simulation. Click on the play button, the rapid forward button, or the single step button once it has reached the post point. Additionally, we have previously addressed this icon during our sessions. Therefore, we should execute this code. Therefore, I will select the "start simulation" icon. Currently, you will observe that this is the interpreter of these activities.

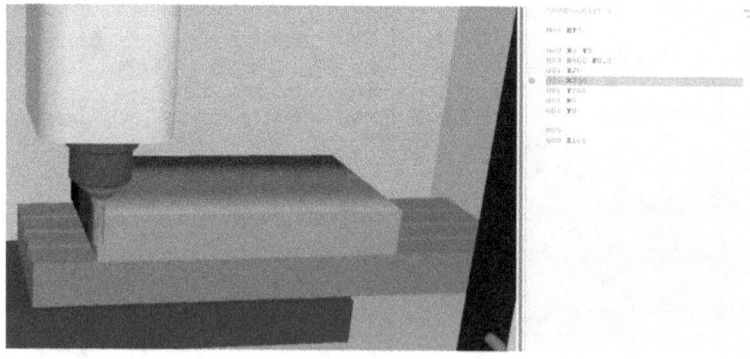

This Simulator Pro automatically reaches the G01X 300 line in the simulation post, and you will be prompted to continue playing in order to continue the G operation. In order to resume the operation as previously discussed, you may utilize the rapid forward button, the single step button, or the play button. Therefore, I will once more select the "play" option. The machining will now commence in accordance with the code that has been provided. Therefore, these post points are exceedingly advantageous when dealing with substantial code.

# HOW TO MAKE COMMANDS

We will engage in a conversation regarding the implementation of comments in the simulator's code. The process of adding remarks to C programs involves two distinct methods. The most prevalent method is to incorporate remarks within the parenthesis. Therefore, let us consider the straightforward program that is presented on the right-hand side of the C and C editor. The code that is written is Z01X 300. While perusing this G01 quarter, it is evident that the linear feed it traversed represents linear movement. Therefore, my objective in composing this document is to relocate the instrument by 300 meters in the x direction. Therefore, I will address this in the comment section. And for that, I will employ these parenthesis. I will write mu at a distance of 300 meters. Therefore, this parenthetical facilitates the inclusion of a comment in the code. Please be advised that the Simulator Pro interpreter will disregard the codes enclosed in the parenthesis. The semicolon command is an additional form of command. It is frequently employed to remark out multiple sections of the block. If I wish to remark out these three lines, I can do so by selecting them in the editor. To select them on the keyboard, you can either hold down the shift key and click the up and down arrows, or you can use the mouse to select them by dragging and holding down the left mouse button. The sections that have been chosen will be indicated by a red highlight.

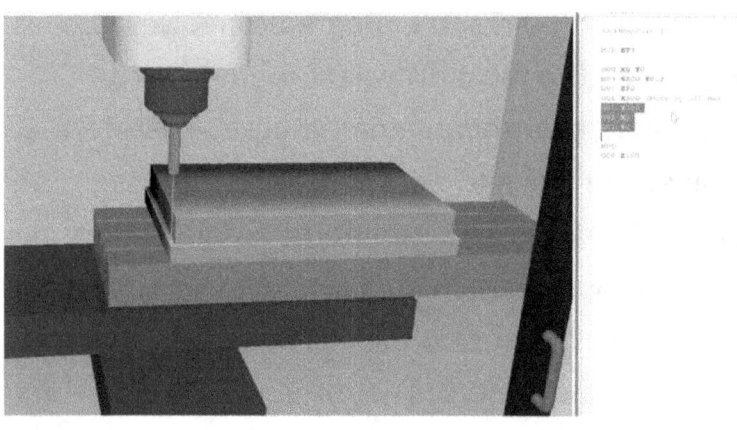

To remark out this selection, navigate to the edit menu and select the "Comment Out" option. Afterward, click on these C and the editor once more. Therefore, it is evident that the hue of this section has been altered as a result of our comments. Additionally, the CNC Simulator Pro interpreter will no longer regard it as an executable command. Therefore, we will attempt to execute this code by treating these lines as remarks. Therefore, I will select this play simulation. Therefore, it is evident that only the operation was executed. That is the tool's movement in the x direction by 300 meters. In order to optimize our perspective, we should rotate this. Additionally, it is evident that the section is machined for D 300 million. Exclusively. Once more, you may select this section to remove the comment. Next, navigate to the edit menu. The option to uncomment is located here. Therefore, choose it. Currently, the CNC simulator is capable of executing this code without any comments. The subsequent critical aspect to comprehend is the process of fabricating these CNC headers. The CNC header can be incorporated into your CNC program by employing the "create consider" function. Additionally, these headers are typically inserted at the beginning of the code to initiate the "Create CNC header" dialog. Select the "Create CNC header" option from the "These Tools" menu. A new overlay window will appear, displaying a selection of various details that can be incorporated into the header. So the initial step is to determine the file name. You are free to specify the name of the file that seems most appropriate. Subsequently, you may furnish the job number, date of code generation operation, and name of the programmer. QSL is the designation we will assign to it. Therefore, it is possible to incorporate the stroke material once more. Regarding our situation. I have chosen aluminum. If you have utilized a machining center, you may furnish its name. The milling center is the specific example we are considering. Additionally, you may furnish the control detail and the material standard measurement that you have chosen. In accordance with your selection, the units are once again mentioned. If you have selected all of these sections prior to initiating the code, they will be automatically populated. Therefore, the information presented here is appropriate for our situation. Therefore, select "okay." The header will be generated automatically upon your selection of "okay." Additionally, you will locate all pertinent information regarding your program here. Therefore, it is recommended that you include the CNC header in the CNC course creation process to gain a more comprehensive understanding of the machine, controls, and workpiece dimensions.

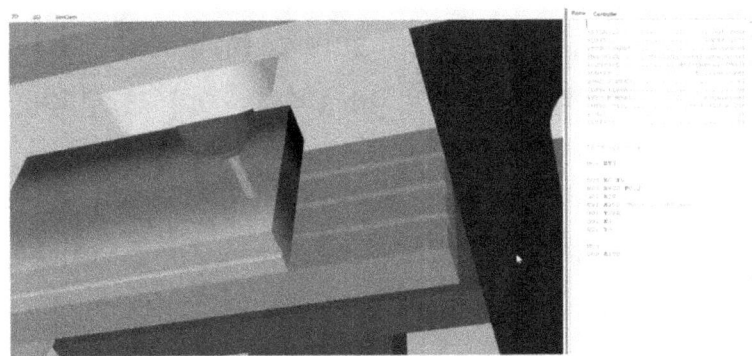

Now, if you make an error while writing the code, such as erasing a significant portion of the program, you can always reverse the change by selecting the control Z key on the keyboard. Suppose that I inadvertently select these. If I delete this, you can revert the action by pressing Control plus Z on your keyboard. Additionally, the Undo and Redo options are accessible from the edit menu. Retrying the action requires selecting Control Plus. Therefore, navigate to the edit menu and locate the Undo and Redo options. Additionally, the shortcut key for both of these options is once again provided in this location. We have previously discussed the basic, yet crucial terminologies that are present in this CNC editor and the methods for utilizing these various functionalities to facilitate the writing of code. The editor can be concealed, allowing the 3D view to encompass the entire window. This is particularly useful when simulating lengthy programs, as the simulation will operate more efficiently with the editor hidden. Additionally, the editor selects the icon located at the bottom of the page, as indicated above. Additionally, the editor or control may be concealed; consequently, I shall opt for this option. Upon selecting it, you will observe that. The entire machine is now accessible from the entire window. Therefore, this is exceedingly advantageous. While the lengthy programs are being executed. The editor or controller can be accessed by clicking on the same icon. The editor can be displayed or concealed by returning to this view menu. Additionally, the Editor or control option is located here. Therefore, choose it. It will be automatically concealed after you have selected it. Once more. Navigate to the view menu and select the editor or controller once more. The controller will be displayed. The editor is once again visible for the purpose of coding.

# INVENTORY BROWSER

In this session, we will address the Inventory browser, the subsequent critical component of this simulator investigation. The inventory browser is the dialog that allows you to generate and maintain all of your virtual workpieces, tools, materials, and zero points. Additionally, it may be referred to as the offset. To access the inventory browser, either click on the settings menu and select the Inventory Browser or enter F2 on your keyboard. The Inventory Browser will be vacant upon its initial launch, necessitating the creation of the necessary items for your simulation. However, it is important to remember that CNC Simulator Pro includes embedded tools, materials, and workpieces for demonstration purposes. This feature is also accessible to the user. At the top of the interface, there are various stages to navigate between the pages in the browser. The initial pages are dedicated to milling and turning tools, followed by workpieces let workpieces material and zero points. Therefore, we will now address the initial entry. In the registry section, that is the Tools section. You have the ability to specify the types of instruments you wish to peruse. The final options are for embedded milling and turning. Therefore, it is referred to as embedded milling tools and embedded tools in this context. To include a new milling tool, select "My Milling Tools" and click the

"Add" icon. Here, you will locate this option. Therefore, I will select it. Upon clicking it. The New Mill Tool editor is initiated, and it is relatively straightforward. Initially, you may furnish the name of the instrument.

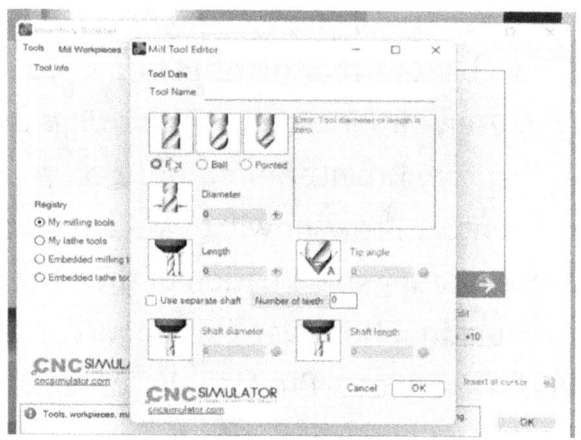

Finally, you have the option to determine the form of instrument you desire. It may be either flat or pointed. Subsequently, you may specify the tool's diameter and length. Once more, you may furnish the soft diameter, soft client, and tip angle in accordance with your specifications. You must select the tool tooltips if you have provided these distinct items so simply. Then, complete the information regarding your instrument and select "okay." However, if you require additional information regarding the configuration, simply select the question mark symbol located to the right of the field. Right now, I will select "cancel." In the same way, you may incorporate the new length instrument to meet your needs. Therefore, it is evident that the blade tool editing is more sophisticated than the email tool editing. Therefore, it is advised that novices refrain from making any modifications to the lead instrument. Alternatively, you may implement the embedded tools in your CNC program. Utilizing the final two sections.

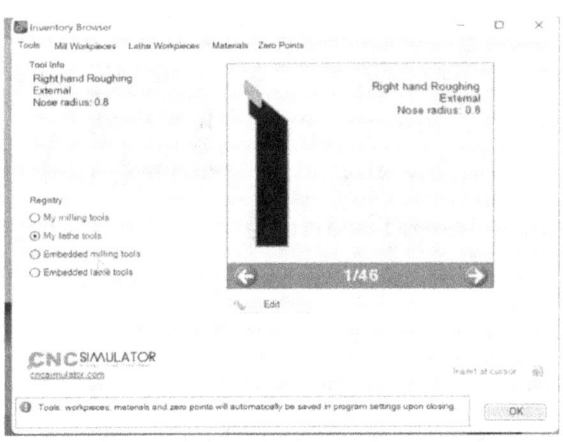

The embedded tools that are included with the CNC simulator Pro can be accessed in the registry section. Use the e t word instead of the standard T code here to invoke a pen embedded tool in your CNC program. Please be advised that this code is specific to c c Simulator Pro and is not a component of the standard G-code language. Therefore, we will proceed to choose these embedded milling instruments. The following is a comprehensive catalog of 68 distinct milling instruments. Additionally, this tool information section furnishes fundamental details regarding each of these instruments upon its selection. Thus, you will observe that if I wish to select the mill with a diameter of 16 and a length of 50, the following will be displayed. First, recall the tool's position. The number 62 is presented below. Therefore, to access this instrument, you must input 80. Then, you must provide the tool's number after you have changed it. If you prefer not to retain the number of this instrument, you may select the option that is indicated by the cursor. Therefore, at present, I shall opt for it. Additionally, you will observe that the board is promptly edited. It is eighty-six. Therefore, it denotes the embedded instrument and its position. That is the 60 instrument that we wish to employ, correct? Now navigate to the preferences menu and select Inventory Browser. Now, if you have incorporated the tools into the Milling Tool section. Additionally, to select the tenth utility, merely enter the letter E. In order to elucidate the selection of this tool to the CNC Simulator Pro, refer to these embedded machining tools once more.

# MILL AND LATHE WORKPIECE

We will address the process of incorporating virtual millwork PCs and network PCs for the purpose of machining. Therefore, to access the budget, navigate to the preferences menu and select Inventory Browser. Proceed to the third stage, which involves the LED for PCs. Consequently, this procedure enables the addition of network PCs. Additionally, the fabrication PCs are located on the second table. Therefore, it is possible to include the necessary information in this location. Therefore, this section also allows you to establish your workstations. Therefore, this is the location where you can include the millwork components that are necessary for the milling process. Additionally, it is important to mention that CNC Simulator Pro includes embedded workstations that are utilized for demonstrations and samples. These are also accessible to users; however, they are not editable or deleteable. To include a new workpiece, select the "add" icon located on the right-hand side. The following is a comprehensive inventory of the various sectors that must be accommodated in light of this size. Consequently, if I wish to have the workpiece's length in the x direction be 150, the lantern's length in the y direction to be 150, and to define the z value, which denotes its profundity, I would press the table again.

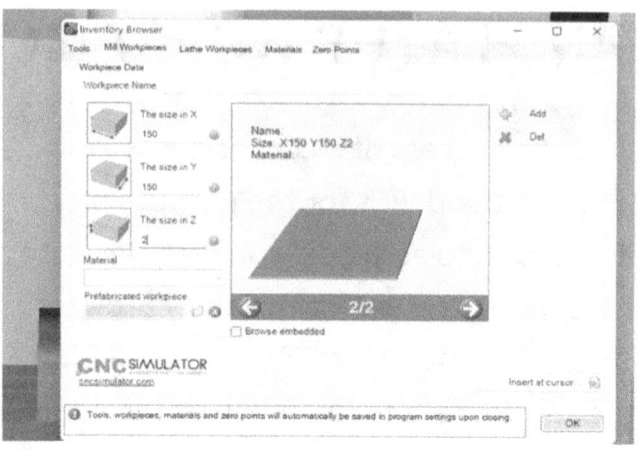

Additionally, if I desire a z depth of 25, I should enter 25. After supplying the dimensions of these workpieces at the top, you may also include the name of the workpiece. Therefore, it is reasonable to assume that this is the masculine workpiece tool. Therefore, I will enter "male workpiece underscore two" and select "OK." Once more, navigate to these settings and select the Inventory browser. Proceed to these Milward PCs. Additionally, the two workpieces are indicated in this location. The initial item is the mill workpiece. One that has been established previously and utilizes this arrow to progress in that direction. So, act on it. Here, you will observe that the mill workpiece has a two-width dimension of 150. One, fifty, and twenty-five. If you have previously defined the material, it will be generated once more by selecting this drop-down menu. Additionally, you may furnish this appropriate material. Therefore, the visual characteristics of the workpiece, including its color and transparency, are determined by the material. Additionally, prefabricated workpieces may be employed. To produce a prefabricated workpiece. It is necessary to develop a CNC program that achieves the desired geometry of the workpiece. This tool menu allows you to rotate the workpiece after the simulation. To do so, select "okay" here. You may now navigate to the menu of these utilities. Additionally, this is where you will locate the alternative. A variety of options, including those that are significant to us, are presented here. That is the only fragment that is being saved. The file can be selected by selecting the yellow folder indicator located to the right of the prefabricated "what is now?" after it has been saved. Once more, navigate to these settings and select Inventory Browser. To navigate through the embedded PCs. Initially, navigate to the Male Workpieces section and then select the Browse Embedded option. Therefore, upon reviewing it, you will observe that the inventory of other inherent components is provided. Subsequently, we shall verify it.

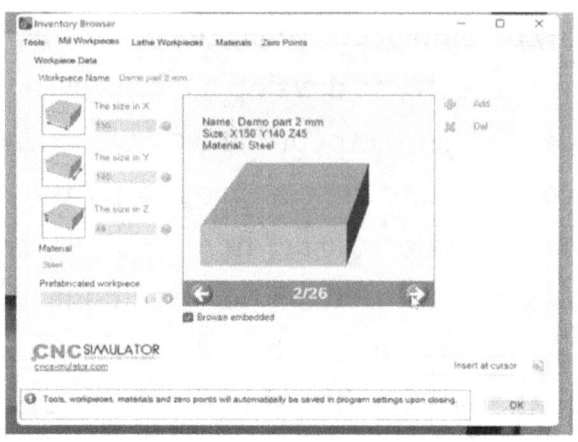

Therefore, upon making the necessary modifications, it is evident that the dimensions and material are also included. Now, if I wish to incorporate these particular workpieces into this program, I can simply select the "insert at the cursor" option. Therefore, I shall opt for it. The dollar aid embedded register is where this index will be written. The portion of the territory. The part number is denoted by these four. Therefore, this is the method by which the milling workpiece can be incorporated. Back to these settings. Navigate to the subsequent phase by selecting the Inventory Browser. That is, eliminate. What features does CNC Simulator Pro include? The embedded workpiece is employed for demonstration and simplicity. The user may also employ these; however, they are not editable or deleteable. To add a new workpiece, simply click on the "add" icon and complete all of the necessary fields for the workpiece's dimensions. Suppose that I wish to have a diameter of 25 and a length of 70. However, you may also assign a name to this workpiece by visiting the "Workpiece Name" section and selecting "Okay." In this section, we have learned how to incorporate the milling workpiece and milling tool into the CNC machine.

# HOW TO ADD MATERIAL

We will deliberate on the process of incorporating additional material into the chosen materials. Define the visual aspect of the workpiece, including color and transparency. Materials may have a single base color and a single top layer color. It is recommended that a thin layer be present at the top, resembling a painted or laminated surface. A method for incorporating the material. Initially, navigate to the inventory browser. To accomplish this, navigate to the settings menu. To add a new material, select the Inventory Browser once more and navigate to the "These Materials" tab. Select the "Add" icon and complete the required fields. Then, employ the color selector to select the desired colors. Check the "Use Top Layer Color" checkbox and select the desired top layer color if you desire a distinct hue. Select the initial code. The initial option contains the material designation, while the metal option is maintained on the right-hand side.

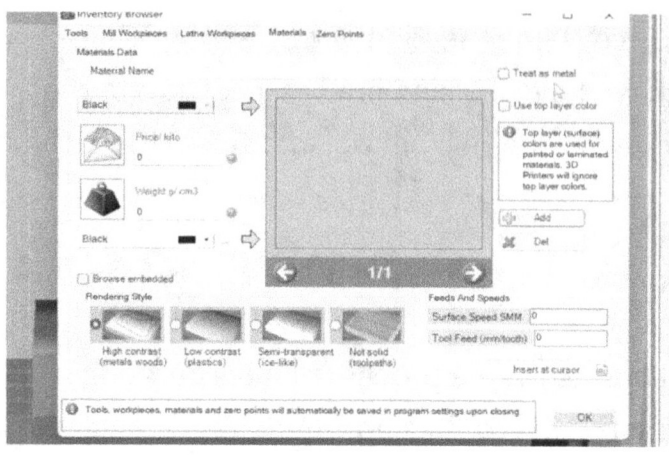

Therefore, the Pretest metal checkbox is presently utilized exclusively by the cutting devices. These devices will emit sparkles, with the exception of the water jet cutter, if they are inspected. The subsequent option at the bottom is the one on the left-hand side. "That is the price or kilo field," which is a material price value utilized by the analysis program. The workpiece weight panel, which can be displayed during the simulation, utilizes the weight field. Selecting the rendering style of your workpiece in the rendering style section allows you to determine how it should appear on the screen. To access embedded materials, select the "browse embedded" checkbox. A selection of various in-built materials will be displayed, and you will be able to determine the month and material by selecting it. Please be advised that materials may be directly inserted into the editor from the inventory browser. Next. I presume that I could incorporate this material by selecting the "insert it cursor" option.

## ZERO POINTS

That is the zero position. The current can be adjusted to 0.2 by utilizing G 92 in the program, and zero points can be accessed from the zero point registry. The message "zero point" is always represented by a dark gray cross on the 3D simulation screen, while the program to zero point is represented by a green cross. Therefore, it is evident in this instance. Let us take a closer look. It is indicated by the green cross in this location. It is important to note that the machine zero point is typically situated in the lower left corner of the table on a milling machine. Zero points, or workpiece offsets, are employed to relocate the origin point. In this C program. You can predefine and invoke 100 offsets in Simulator Pro from your programs. The G 54 to Z 59 code utilizes the first six of them, which are indexed 0 to 5. Once more, navigate to these settings and the Inventory Browser.

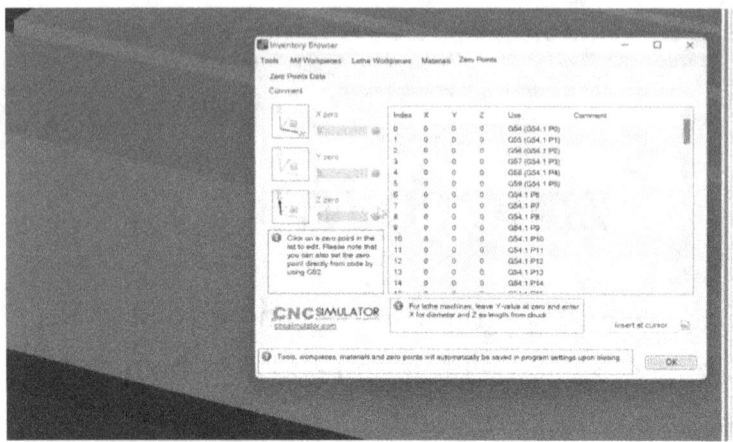

Choose a zero position at this location. To select zero Point Offset number three from your program, merely enter G 57. The command G 54.1 is used to select an offset with a higher index than five, as demonstrated in this example. Additionally, it is important to mention that it is possible to directly incorporate a zero point from the inventory browser into the editor by selecting it and choosing the "Insert Add cursor" option. Therefore, it is presented here. The "Insert cursor" option is accessible.

# BASICS OF G CODE PART 1

The syntax and structure of the key codes will be the subject of our discussion. Finally, we will utilize the G codes to address a few of the examples. And the two are maintained. We will commence with the G code. Those are the G00, which denotes the repetitive auditioning. G00 is the rapid positioning code that is employed to expedite all movements to a designated point without the need for machining. Therefore, it is evident that this G00 code is only beneficial for quicker movements and does not generate any machines. It is frequently implemented to expedite the tool's transition between machining operations or to relocate it away from the workpiece. The machine operates at its utmost rapid rate, as G00 does not specify the O input rate. The syntax of this G00 will now be the subject of discussion. To employ this G00 code. The initial step is to compose the letters DG00.

## G00 - Rapid Positioning

Syntax:
G00 X__ Y__ Z__    (or other relevant axes)
Here, X, Y, and Z are the coordinates where the tool should be positioned.

Then, you can specify the appropriate axis for the movement, which may be the x, y, or z axis, or a combination of them. Note that the coordinates x, y, and z are the locations where the tool should be positioned. Therefore, if you wish to relocate 20 meters in the x direction, what is the equivalent distance in the y direction and an additional m in the z direction? Subsequently, you will implement. This code is formed by the characters G00, x 20, y 40, and z. It is no longer necessary to utilize all of these axes in accordance with your instrument movement. The coordinates for this particular axis may be specified. The linear interpolation, which is coded as g zero, is the second most critical instrument. Consequently, these G01 are the linear interpolation G-codes that are employed to direct the Z tool to move in a straight line from one point to another and to regulate it. It is crucial to develop a straight line tool element for operations such as drilling, profiling, and contouring. This zero one allows you to specify a transmission rate that will determine the speed at which individuals should move between the points. Therefore, the G0 one tool can be utilized to facilitate the movement of the object in a straight line for any desired length. Additionally, it is important to mention that the G01 code is employed to perform the machining. Additionally, you may establish a feed rate that corresponds to your specifications in order to conduct the machining process once more. The syntax for this G0 one will now be the subject of discussion. In the same way, this G0 one can be employed. Initially, it is necessary to compose DT1. Then, in accordance with your specifications, you may furnish the coordinates of the d, x, y, and z axes.

## G01 - Linear Interpolation

Syntax:
G01 X__ Y__ Z__ F__ (feed rate)

X, Y, and Z are the coordinates of the target point.
- F__ is the desired feed rate in units per minute (e.g., inches per minute or millimetres per minute).

Finally, the feed rate at which the machining is to be conducted can be specified. The coordinates of the target point means are represented by x, y, and z. Suppose that we wish to transition from point 1 to 2.2. It is evident that the instrument will be accessible at 3.1. Therefore, in order to relocate to point two, it is necessary to furnish the coordinates of the second location. It may consist of any combination of right coordinates, including x, y, and z. The desired transmission rate, denoted as f, is expressed in units of minutes. That is, if we have utilized each unit, it will be regarded as the h per unit. Alternatively, if you have employed the millimeters unit, it will be interpreted as millimeters per minute. So in conclusion, the G00 code is employed for rapid movement in cases where the machining is not flawless, while the G01 code is employed for straight-line movements. This straight line machining is executed by employing this code.

# BASICS OF G CODE PART 2

Our discussion will commence with DG02 for G02. The circular interpolate action is primarily performed in the clockwise direction using code. Therefore, G02 is the clockwise direction. The G-code for circular interpolation, which is employed to generate walkways, spans four seconds. It specifies the arc's center radius and endpoints.

G02 is frequently implemented to generate circular or rounded corners. Let us deliberate on its syntax. Therefore, in order to employ these g two codes, it is necessary to first specify d, G02, and subsequently provide d coordinates. That could be the x-coordinate or the y-coordinate. The endpoint coordinates of the are represented by x and y, and the endpoint is then specified. d, I, and j are coordinates that specify the center of d, I, or and can be employed. Alternatively, the guide may be employed to determine the radius of the arc.

## G02 - Circular Interpolation (Clockwise)

Syntax:
G02 X__ Y__ I__ J__ (or R__)
- X, Y are the endpoint coordinates of the arc.
- I, J specify the coordinates of the center of the arc.
- Alternatively, you can use R__ to specify the radius of the arc.

Therefore, in order to comprehend this, we will provide an example. Suppose that I wish to create a circular cut from one point to d, another point in the clockwise direction. At the end point, the radius is ten. In this case, I can simply provide the coordinates of x and y from 0 to

10, as well as the radius c of ten. Currently, the center point of the region can be specified in addition to the radius, which is represented by the letters I and j. Therefore, in the majority of instances, we will employ the radius r to denote circular interpolation. The subsequent significant code is DG034. Again, it serves as an indication of this circular interpolation. G03 is counter-clockwise, as it is advantageous for counterclockwise rotation.

### G03 - Circular Interpolation (Counter clockwise)

Syntax:
G03 X__ Y__ I__ J__ (or R__)
- X, Y are the endpoint coordinates of the arc.
- I, J specify the coordinates of the center of the arc.
- Alternatively, you can use R__ to specify the radius of the arc.

The circular interpolation code is employed to generate counterclockwise arcs in a manner similar to G02. It specifies the center radius and end locations of the arc. G03 is frequently employed to generate circular contours or rounded corners in the opposite direction of G zero. The index of using these g zeros to be virtuous is once again the subject of discussion. Therefore, in G03, it is necessary to initially provide the coordinates of the arc's endpoint, which may be either the x coordinate or the y coordinate. This is followed by the specification of d, I, and j, or the same method can be used to specify the radius. The coordinates of the earth's center are denoted by I and j, respectively. Alternatively, the radius of the arc is denoted by R. Therefore, in conclusion, G02 and G03 are considered. The syntax of both is identical, and they are exclusively employed for circular interpolation. However, in order to achieve a clockwise movement, the G02 chord must be employed. Additionally, the G03 must be employed to achieve either the anti-clockwise or clockwise movement at that moment. Regarding these. For circular contours or rounded corners, both are implemented. The G04 code, also known as the well, is the next significant code. It is the dwell code that is used to post the machine for a specified duration at a specific point in the program. This command is beneficial for operations that necessitate a host, such as ensuring that a tool chain is complete or permitting coolant to circulate. Let us deliberate on the syntax of this command. In order to employ this code, it is necessary to first specify DG0, and then the p must be specified. Therefore, this p denotes the weight. In other words, subsequent to the letter "p." For instance, if you wish to maintain the current value for one, two, or three seconds, or s, but your requirement is to provide d time in a, d g after recording the p. Therefore, the initial critical G codes are as follows: d, g, double zero, G01, zero two, zero three, and g zero for all. These codes will be extremely beneficial in any of these machining operations.

# BASICS OF G CODE PART 3

We will commence our discourse with the number five. What is G05? Code, which is also referred to as the HSM, is a form of high-speed machining. So these G05 activities are high-speed machining. Additionally, it is implemented when CNC machining is equipped with high-speed machining capabilities, which facilitates a more efficient and smoother process, but diffusion. Therefore, in order to employ this code, one must merely type G05. The polar coordinate is indicated by the code d g 16, which is executed prior to the commencement of any subsequent critical code. Therefore, the G code g 16 is employed to facilitate polar coordinate programming. Total movement is the fundamental method of employing polar coordinates in polar coordinate mode, as opposed to the conventional Cartesian coordinates. Therefore, the distance ahead is denoted by polar coordinates. In order to comprehend this, let us consider an example. Assume that the instrument is accessible and this particular point serves as the center. Our current goal is to ascend to the highest point of this specific circle. Therefore, in order to achieve this particular circular movement, we are aware that the provisional angle and radius will be more advantageous than the Cartesian coordinates. Therefore, it is sufficient to furnish the coordinates of these and the object. Then, you must specify the perspective from which you wish to approach the matter.

### G16 - Polar Coordinate Programming

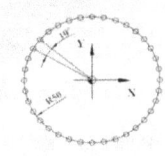

- Polar coordinate programming mode.
- Tool movement is specified using polar coordinates (distance and angle)

Therefore, these polar coordinate programming would be necessary if you wish to utilize this specific angle and radius for the tool's movement. Naturally, these 16 enable the machining operations that would be performed on entire circular or angular portions. Therefore, these are precise coordinates. This term is beneficial whenever we encounter angular movements during our machining operations. The syntax of these 16 codes merely signifies that the polar coordinates system is activated by typing G 16. Afterward, we will only provide the distance and angle as the coordinate system when we conduct any machining operation with that. The rectangular coordinate is the subsequent critical instrument. The subsequent critical command is d g 15. It also suggests that the decoder used to return to the rectangular coordinate program is G 15. Following the implementation of the polar coordinate programming. Therefore, the polar coordinate programming is activated if we have employed d g 16. And if we wish to return to the rectangular coordinate program at that time, we will again employ these fifteen. Therefore, this code reverts the CNC machine to the conventional Cartesian format for interpreting coordinates. That is, the coordinates of x, y, and z. In comparison to other sequences, D 15 and 16 are utilized less frequently. However, it may prove advantageous in scenarios where circular or angular movements necessitate more intuitive programming. Utilizing polar coordinates to enable CNC programming for particular applications. In conclusion, g 15 is employed to leverage Cartesian coordinates. Okay 16 is employed to implement d polar coordinates. 15 is only helpful when the polar coordinate system is activated. Therefore, it is advantageous to return to the rectangular coordinate programming of D 15. During our subsequent sessions, we will address the G codes' D.

# BASICS OF G CODE PART 4

We will commence our discourse with the number five. What is G05? Code, which is also referred to as the HSM, is a form of high-speed machining. So these G05 activities are high-speed machining. Additionally, it is implemented when CNC machining is equipped with high-speed machining capabilities, which facilitates a more efficient and smoother process, but diffusion. Therefore, in order to employ this code, one must merely type G05. The polar coordinate is indicated by the code d g 16, which is executed prior to the commencement of any subsequent critical code. Therefore, the G code g 16 is employed to facilitate polar coordinate programming. Total movement is the fundamental method of employing polar coordinates in polar coordinate mode, as opposed to the conventional Cartesian coordinates. Therefore, the distance ahead is denoted by polar coordinates. In order to comprehend this, let us consider an example. Assume that the instrument is accessible and this particular point serves as the center. Our current goal is to ascend to the highest point of this specific circle. Therefore, in order to achieve this particular circular movement, we are aware that the provisional angle and radius will be more advantageous than the Cartesian coordinates. Therefore, it is sufficient to furnish the coordinates of these and the object. Then, you must specify the perspective from which you wish to approach the matter.

## G16 - Polar Coordinate Programming

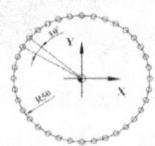

- Polar coordinate programming mode.
- Tool movement is specified using polar coordinates (distance and angle)

Therefore, these polar coordinate programming would be necessary if you wish to utilize this specific angle and radius for the tool's movement. Naturally, these 16 enable the machining operations that would be performed on entire circular or angular portions. Therefore, these are precise coordinates. This term is beneficial whenever we encounter angular movements during our machining operations. The syntax of these 16 codes merely signifies that the polar coordinates system is activated by typing G 16. Afterward, we will only provide the distance and angle as the coordinate system when we conduct any machining operation with that. The rectangular coordinate is the subsequent critical instrument. The subsequent critical command is d g 15. It also suggests that the decoder used to return to the rectangular coordinate program is G 15. Following the implementation of the polar coordinate programming. Therefore, the polar coordinate programming is activated if we have employed d g 16. And if we wish to return to the rectangular coordinate program at that time, we will again employ these fifteen. Therefore, this code reverts the CNC machine to the conventional Cartesian format for interpreting coordinates. That is, the coordinates of x, y, and z. In comparison to other sequences, D 15 and 16 are utilized less frequently. However, it may prove advantageous in scenarios where circular or angular movements necessitate more intuitive programming. Utilizing polar coordinates to enable CNC programming for particular applications. In conclusion, g 15 is employed to leverage Cartesian coordinates. Okay 16 is employed to implement d polar coordinates. 15 is only helpful when the polar coordinate system is activated. Therefore, it is advantageous to return to the rectangular coordinate programming of D 15. During our subsequent sessions, we will address the G codes' D.

# BASICS OF G CODE PART 5

Our discourse will commence with the subsequent G chord. This is the G17 chord, which denotes the x- and y-axes. Therefore, the x-y plane is the plane of operation for these G17. Additionally, it specifies that all subsequent movements and calculations will be conducted in the x-plane. Therefore, we will employ these inputs exclusively when we are required to execute the movement in the x-y plane. And to utilize these, you must simply input G17 before composing any of the other modes. Therefore, the x and y coordinates will be the only ones taken into account when the coordinate points are specified. Therefore, this g can be employed as the syntax. Here, it is evident that the coordinate system of d, x, y, and z has been provided. Therefore, we will employ these only if we wish to execute the movement in the x- and y-axes. These. The G chord is the next critical chord, and it signifies the selection of the Z plane. G 18 selects the x-z plane as the plane of operation. It specifies that all subsequent movements and calculations will be performed in the Z. Therefore, if our objective is to obtain the entire movement in the Z plane at that moment, we will employ these g inputs. In order to implement these eight in code.

# G18 - XZ Plane Selection

- XZ plane as the plane of operation.
- Subsequent movements and calculations

The code must be preceded by the expression "d g 18." Therefore, all movements will be executed in the exit plan once you specify the residual 18 g. The subsequent G code is G 90, which signifies that the Y plane has been selected. These G 19 designate the y-z plane as the plane of operation, indicating that all subsequent movements and calculations will be performed in the device plane. Similarly, to employ this G 19 code, merely enter "g 19" into the editor. The machining operation will only take into account the residual coordinates G, y, and z after you enter g 19. The G codes that have been discussed thus far are the foundational elements of CNC machining and were instrumental in the regulation of tool movements. During the machining operation, the selection of the plane and the pause or dwelling are specified. The proper comprehension and application of these D codes are indispensable for CNC programmers and operators. In our subsequent session, we will address several additional significant G codes.

# BASICS OF G CODE PART 6

We will commence our discussion with the subsequent significant code, which is 20. These G 20 are beneficial for the. Each unit. The G code G 20 is employed to configure the machine to operate in inches. It is the Imperial unit. All length and coordinate values in the program are interpreted in the presence of G 20. This is the case whenever your component has a dimension in the form of. Initially, we must activate the G 20 code. To do so, merely enter the code "G 20." Prior to composing the code. Therefore, the coordinates that you have specified will be regarded as inputs once you have included g 20 in the code. The G 21 command is the subsequent critical command. Additionally, it specifies the metric units. The G code G 21 is employed to configure the machine to operate in millimeters. Units that are asymmetric. All land and coordinate values in the program are interpreted in millimeters when G 21 is active. Therefore, it is necessary to employ the g 21 command whenever your component contains metric inputs, which are indeed millimeter dimensions.

# G21 - Metric Units

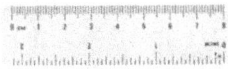

Syntax:

G 21

After composing any of the coordinates that will be taken into account in the millimeter inputs, you utilize them as a syntax. You simply need to enter D, g 20. The g 28 is located immediately adjacent to the g code. Additionally, it signifies the return to one's residence. The G code G 28 is employed to relocate the tool to the machine's home position or predetermined reference points. It is frequently employed in the context of tool modifications or as an element of the assistance process. The coordinate value for x, y, and z is typically present in these G 28. In other words, the d zero coordinates of x, y, and z are defaulted to the x zero, y zero, and z zero. Additionally, you may designate the precise locations in accordance with your machine. The G 90 is the subsequent g code, which denotes the absolute program.

Therefore, these G90 are the G codes that are employed to configure the machine for absolute programming in absolute mode. The program interprets all subsequent coordinates as absolute positions from a predefined reference point, which is typically 00G 90. This ensures that the coordinates represent exact positions in the machine's coordinate system. Utilize the following 90 code syntax. You need only enter G 90, and the program will recognize all of the points as the absolute points after you input any of the coordinates.

## G90 - Absolute Programming

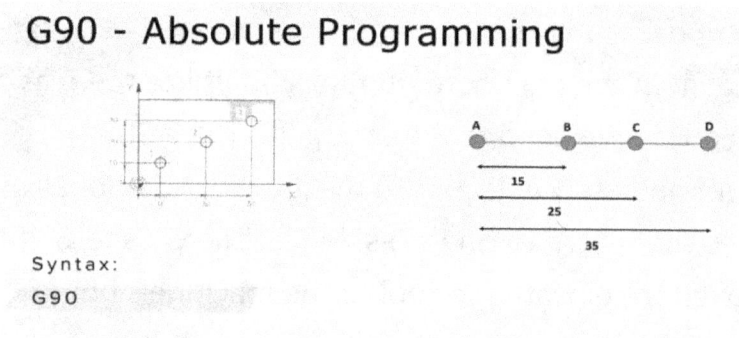

Syntax:
G90

In order to comprehend imperative programming, we will employ this example. The line data on this screen indicates four points: A, B, C, and D. It is assumed that point A is the machine zero or the zero point. In order to utilize these absolute programming techniques, it is necessary to supply the coordinates for points B, C, and D solely from these reference points. So, as you can see, the distance between B and point A is 15 units. Therefore, you must relocate to this point B and provide the x coordinate 15. In the same vein, the C. is the most direct route to the objective. Therefore, the distance from B to C cannot be specified when the g 90 option is selected. However, the distance from point A to point C must be specified, and it is D 25 units. Therefore, in order to access this location, you must supply the x coordinate value. That is d 25. In the same way, in order to proceed to this point D, you must once again specify the distance from each reference location. That is the eighth point. Therefore, in order to advance to this point D, you must supply 35 units. So as you can see, we are not providing the distance from the last point; rather, we are providing the coordinate point from the reference point or d zero, which provides the d dimension. These absolute programming modes are available for your utilization. Additionally, it is necessary to select e G 90. Come on. The subsequent significant G code is D g 91. It also indicates that the incremental G 91 is the G code that is used to configure the machine for incremental programming in incremental mode. The incremental movements are interpreted as all subsequent coordinates. Twenty instruments. The present position D 91 is beneficial when determining the amount by which the instrument should be moved in relation to its current position. Utilize these incremental programming techniques. Syntax is straightforward.

## G91 - Incremental Programming

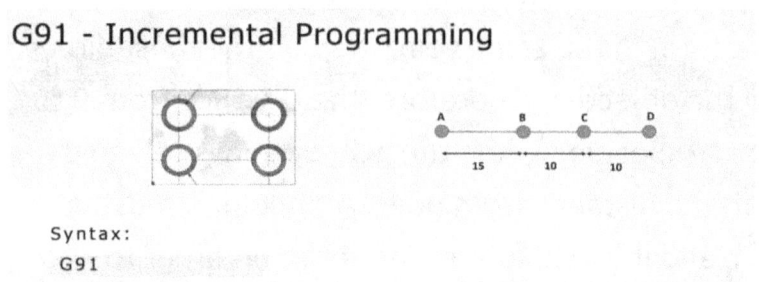

Syntax:
G91

In other words, you must input G 91 into the editor. Once more, in order to comprehend these incremental programming techniques, I will examine the 2D elements that are displayed on the screen. Next, let us assume that point A is our zero or reference point. The x coordinate value must be provided in order to proceed to point B. That is, the 15 units from point B to point to determine whether you wish to relocate. Then, if you have chosen G 91, you must merely specify the distance of the point a c from the current tool position. Therefore, in order to proceed to this point, you must merely supply D ten units in the X direction, as the instrument is already accessible at point b. In the same way, to transition from point C to D, you must once again provide ten units, as the instrument is exclusively available at point C. Therefore, as you can observe, we are not following the distance from the reference point; rather, we are supplying the distance of the subsequent point from the present tool position. Therefore, it is necessary to employ this G 91 in order to provide d dimension in this incremental mode. In conclusion, both G 90 and 91 are beneficial in determining the distance to the next available point. It may be in the absolute mode or the incremental mode. It is advisable to employ these G 90 if you wish to provide the distance for the entire component from a single reference location. That is the ultimate programming. However, incremental programming will be extremely beneficial if you wish to specify the distance between the present position of the tool and each component. Therefore, I trust that you have gained an understanding of the significance of these two critical programming tools, which are used to instruct course 90 and 91. The following significant key elements will be the subject of our subsequent session.

# BASICS OF G CODE PART 7

Our discourse will commence with the subsequent critical G-code, which is the GMAT. It is a genuine measurement that indicates the number of feet per minute. The G-code 1894 is employed to establish the feet rate in units per minute. It specifies that the feet rate in the program denotes the distance the instrument travels per minute. Therefore, the pace per minute is crucial when you provide the distance movement at 42 meters per minute and the corresponding duration. And in order to accomplish this, you must choose the G 94. G 94 is frequently employed to regulate supply rates. Machining procedures. The syntax for this G 94 is straightforward: merely enter G 94. Therefore, the G 94 command is executed by a controller, which automatically recognizes that the feed you have supplied must be taken into account when calculating the feed per minute. The G 95 is the subsequent instrument, which is employed to evaluate the input per revolution. The feed rate is determined in units per spindle using the G-code G 95. It specifies that the input rate in the program represents the distance between the majority of individuals. G 95 is frequently employed for threading and twisting. The syntax for these G 95 commands is as simple as typing them. Nevertheless.

## G95 - Feed Per Revolution

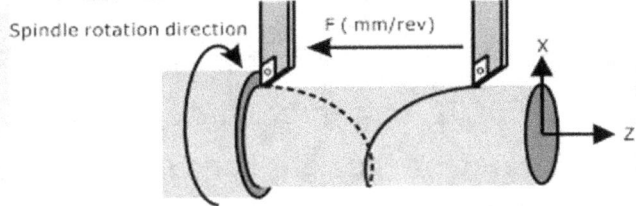

- Feed rate in units per spindle revolution.
- Distance the tool moves per spindle revolution.

Therefore, whenever a controller executes the G 95 command, it. It is important to note that the feed you have supplied must be taken into account when calculating the feed per revolution. Therefore, the G 94 or 90 can be implemented in accordance with your specifications and the machining operation. It is crucial to comprehend the appropriate duration to employ these G 94 and the utmost rotary operation in order to complete the machining on the rotating component. At that time, it is necessary to supply power to the LED machine, which may be a threading or turning operation. However, for the linear cut or the computer's construction, it is more advantageous to employ the G 94. That is the utilization of supply per minute. I trust that you have a clear understanding of the appropriate use of the G 94 and G 95 codes. We will examine specific videos in order to comprehend how to utilize them in our subsequent decision. A few of the most significant G reports.

# SIMULATION OF G CODE PART 1

We will comprehend the utilization of several critical codes, including T000102, zero three, and zero four. Therefore, as illustrated on the right-hand side of the screen, we have generated a straightforward code. Additionally, it is evident that G00 has been incorporated into this code. Therefore, this G00 is advantageous for swift movement. The coordinate points that are EX0 and y zero are provided here. Subsequently, you will observe that we have furnished coordinates for z 20 in TG01. The linear interpolation is denoted by G01, and the z value of 20 indicates that the tool will relocate from the reference position to the 20. The following item is the x 300. Therefore, it suggests that the tool will move in the x direction up to 300 during the machining process. The Y 200 is denoted in the subsequent line. This implies that the instrument will proceed in the y direction while maintaining the x direction as it was in the previous command. We are not required to provide the x coordinate after we have supplied it in the subsequent line.

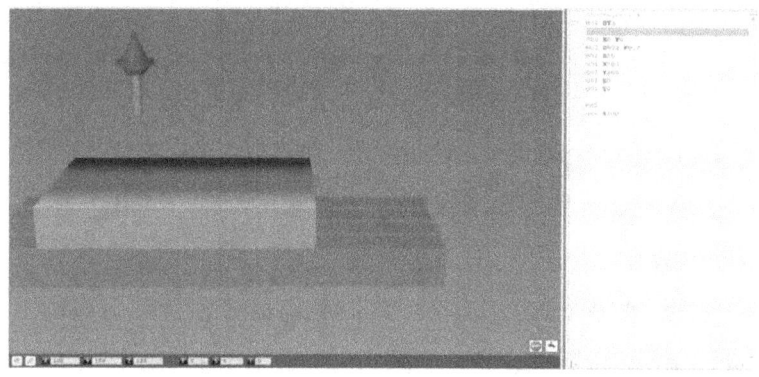

Once more, the interpreter will automatically interpret the x coordinate as 300. And it will now modify the y-coordinate. In other words, the 200. Therefore, the instrument will progress from 0 to 200. Therefore, the coordinate has been reset to zero in this instance. Therefore, the y-coordinate will be regarded as 200. Additionally, the zero coordinate is regarded as the x coordinate. Therefore, the instrument will transition from the x 300 coordinate to the zero coordinate. Finally, y zero is supplied. Therefore, the x coordinate will be assumed to be negative. Additionally, the y coordinate is again treated as a zero. Therefore, the instrument will transition from y 200 to y 0. Finally, we have introduced the G00Z 100, which enables the tool to travel to a distance of 100 MB from the reference point without the need for machining. We are already aware that the green cross mark is the reference location for the 41 piece. As is evident at the outset of the component. Therefore, the reference zero location is indicated by the green cross mark. In the event that you have not supplied any incremental or absolute programming code, the interpreter will default to treating it as an absolute program. In order to comprehend this code, we will attempt to execute it. Therefore, examine the right-hand side. Additionally, the position of these gray lines is adjusted proportionately. The tool's movement is visible on the left-hand side. Subsequently, the operation is multiplied by 300. Consequently, y 200 is once more equal to zero. Presently, the value of x is negative.

Lastly, the utility will transition from the reference position to a distance of 100 units from the reference point. I will replay this video in order to acquire a more comprehensive understanding of the subject matter. It will appear as intended. I trust that you have a clear understanding of how to utilize this command and how to input the coordinates into the program. We have employed a few of the M commands, and we will elaborate on the M command in our subsequent session. I will once more present you with a selection of valuable videos for G02, zero three, and G04.

# SIMULATION OF G CODE PART 2

I will demonstrate the operation of G02, zero three, and zero four. Therefore, the animation area where the machining is conducted will be visible on the left-hand side of the screen. The command is located on the right-hand side, and we have specified the coordinates in accordance with our specifications. Initially, examine the right-hand side of this page. Initially, we have chosen EG00. Therefore, it suggests that the positioning is accomplished rapidly. Additionally, the x coordinate is zero, while the y coordinate is 200. You are now invited to descend once more and examine the G01 command. Additionally, we have specified that the z coordinate is d 20. Subsequently, we have implemented anti-tool movement for the 215 units in the x direction. Next, you will observe the DG02 command, which denotes a circular movement in a clockwise direction. Therefore, we are furnishing the x coordinates, which are 300, and the y coordinates, which are 150. And in order to achieve these two specific end points, we are specifying a radius of d 50.

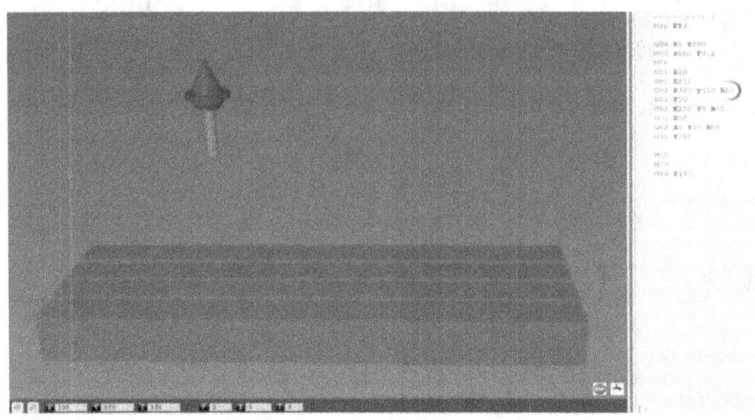

Therefore, the radius of 50 will be established from the final point and extend to the end point. That is, the x coordinate is 300 and the y coordinate is 150. Similarly, this code can be employed to comprehend the various coordinate points that we have supplied. Initially, I will execute this code in order to gain a more comprehensive understanding of the written code. In an effort to comprehend the movements, I shall ensure that I observe the screen effectively. Thus, as you can observe, we are initially shifting the bottom in the Z direction. Subsequently, we have furnished. Is a direction present? Subsequently, we will proceed in the direction of x. We are once again offering the circular contour curve. Additionally, it is evident that it is in a clockwise orientation. Therefore, we will implement these. These two. Once more, we are progressing in the direction of x. Therefore, we have furnished the y coordinates. We are now presenting this circular interpolation in a clockwise direction once more. Therefore, g two is once more employed to denote linear interpolation. The x coordinates are furnished. Lastly, this circular interpolation is provided in the clockwise direction once more. Therefore, the G02 command is implemented. And finally, the y coordinates are provided to return to the home position. I will re-run this video to ensure that you have a more comprehensive understanding. In accordance with the coordinates that were supplied. We will now proceed to the subsequent video. First, I will play this video to provide you with a more comprehensive understanding of the machining process. Therefore, examine the display.

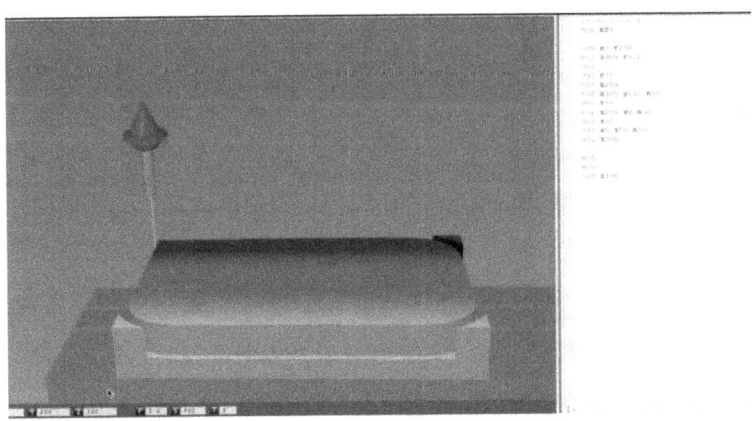

Therefore, it is evident that we are initially proceeding in the x direction. Therefore, the g g0 one code has been employed to provide the x coordinate. We are now presenting the circular contour in the anticlockwise direction at the 250 x coordinate. Therefore, we have implemented G0 three. The circular interpolation, which is 300 x, is represented by the x coordinates that serve as the termination points of these. Additionally, divise fifty. Additionally, we intend to establish a radius of 50. Therefore, we have referenced r 50. We are finally returning to our original position. Therefore, the initial step is to set the value of x to zero. Subsequently, we achieve a zero value for x. In order to relocate the tool from the workpiece, we must employ the G00 command and the t z coordinate 200. I will once more play this video to provide you with a more comprehensive understanding. Please examine this display. In these two videos, we have gained an understanding of the distinction between the G02 and zero three commands and the impact they will have. We will now proceed to the subsequent video. Additionally, the G01 command is explicitly implemented in this video. In the same vein, T03 and four. On the right-hand side, you will observe that we have implemented the G04 command, which denotes a boost time of 44. So, the G zero for P 3000 has been referenced in this context.

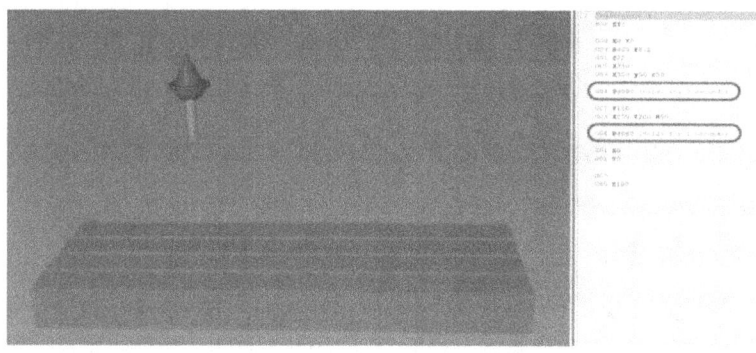

Therefore, these seconds are denoted in milliseconds. Therefore, the machining will cease for 43 seconds as indicated by these 3000. Therefore, we should play this video. It will now remain in place for three seconds. The machine is once again in operation after being held for three seconds. So, I trust that you have grasped the concept that G04 is extremely beneficial when it comes to maintaining the machining operation, the tool chain, or the coolant chilling. Therefore, it is crucial to comprehend the significance of these. Some of the fundamental G codes that are used to execute the machining process in this CNC control. It is now apparent to us what the purpose is.

# BASICS OF M CODE PART 1

The M codes will be the subject of discussion. M codes are auxiliary codes that are an essential component of CNC programming. They are accountable for the supervision of a variety of machine functions and auxiliary operations that occur during the machining process. Therefore, this module will address the most frequently employed M codes. The syntax and structure of M codes will be the subject of our subsequent discussion. Lastly, the function of various M codes in regulating the spindle refrigerant and auxiliary function. The initial m code is M00, which denotes the program scope or optional. Therefore, M00 is the M code that is employed to momentarily halt the CNC program. It is frequently employed for manual intervention, which enables the operator to halt the program. Therefore, these M00 are extremely beneficial whenever we need to manually and momentarily suspend the program. The machine remains in a halt state until the operator expressly resumes the process while applying this code. Therefore, the user is required to restart the program in order to utilize the M00 code after it has been applied. To put it simply, you must input M00 into the program. The program is automatically terminated by the interpreter upon the execution of the M00. Now, in order to restart the program, you must initiate it manually. The subsequent critical M code is the optional program halt, which is also referred to as the conditional stop. The M01 is a conditional halt, but it is analogous to the M00. The program was typically paused only when specific conditions were fulfilled.

M01 - Optional Program Stop (Conditional Stop)
- Conditional stop.
- Pause the program only if certain conditions are met.

The distinction between M00 and M01 is straightforward, as operators have the option to continue the program execution after evaluating the condition. At any time. If our goal is to proceed to the program after satisfying specific criteria, we will employ the M01 code at that time. Additionally, if you wish to terminate the program during execution, the code M00 will be inserted. To utilize this command effortlessly, input DM01 into the program editor. The subsequent m code, which is the M02, is beneficial for terminating the program. The M code M02 is employed to denote the conclusion of the CNC program. The program execution is halted when the machine encounters M02, and it typically returns to the program's inception. M02 is frequently implemented at the conclusion of a CNC program. Therefore, it is more expedient and effective to employ these M02 codes at the conclusion of the program, as they entirely close the program. Consequently, the initiative is terminated and the procedure is finalized. To reactivate this code, simply enter M02 into the program editor. The M03 is the next most critical command. Additionally, it is advantageous to activate the sprinter and conduct measurements at an early stage with these M03. This M03 is utilized to initiate the spindle rotation integral quiet detection, which is accomplished by turning on the spindle in a clockwise direction. It enables the spindle to rotate in a predetermined direction by activating it. M03 is frequently followed by a s code to establish the spindle speed. To utilize this command in a straightforward manner, the M03 must be entered. After that, you will enter S and a specific unit of measurement to represent the spindle's speed.

## M03 - Spindle On, Clockwise

Syntax:
M03

Therefore, if you write M03S 1000, the spindle speed is denoted by 1000. In the same manner. The subsequent critical command is "m." That is the M04. Additionally, it is employed to activate the spindle in a counterclockwise orientation. Therefore, these M04 are comparable to these M03. However, this code activates the spindle in the opposite direction of the M03 code to initiate the counter-clockwise rotation. Additionally, the M04 code is followed by a S code to establish the spindle speed. To execute this command, merely enter M04, followed by the spindle speed, which is assumed to be 1000. Subsequently, you will input M04S 1000. In this case, 1000 denotes that the spindle will rotate at a speed of 3000, while M04 denotes that the spindle will rotate in a counterclockwise direction. The M03 and M04 are both beneficial for turning on the spindle, as we are aware. However, if we wish to activate the spindle in a clockwise direction, M03 is advantageous. And if we wish to activate the spindle in a counterclockwise direction, M04 is advantageous. In the same vein, the M05 is the subsequent critical code. Additionally, it is employed to deactivate the spindle. Therefore, the M05 code is employed to halt the spindle rotation. It brings the spindle to a halt by deactivating it. At the conclusion of machining operations or when a tool change is necessary, Condition M05 is frequently implemented. It is important to observe that these M05 are frequently employed. To utilize this code, it is sufficient to execute the M05 command after having terminated the spindle with M03 or M04. The spindle will be automatically deactivated when the interpreter executes the command M05 in the program. Therefore, I trust that you have gained an understanding of the primary significance of M force. So far, we have addressed the most critical aspect of the M code, which is the process of activating the spindle during machining. This code enables the automatic activation and deactivation of the spindle rotation in a similar manner.

# BASICS OF M CODE PART 2

We will commence our discussion with the middle six, which is the next significant M code. The M06 code is utilized to initiate a tool change process, enabling the machine to transition to a different tool while machining. Therefore, this particular command is extremely beneficial when trimming intricate geometry. And as we employ various tools at specific intervals during the machining process, these M06 are followed by a tool number and other essential parameters to facilitate the utilization of such tools. To employ this code, first enter M06, followed by the n t code. Your input will be the tool number. Therefore, the tool selection is denoted by T. Subsequently, the tool that is selected from the available tools will be the specific number that was entered. So, if I wish to utilize the drill tool and it is accessible at the second position, we will enter the command as M0602.

## M06 - Tool Change

Syntax:
M06 T__ (Other necessary parameters)

T__ indicates tool number

Therefore, the M06 interpreter will automatically recognize that it is time to replace the instrument and determine the appropriate moment to do so. Then, that number is zero. The instrument change number will be taken into account, in addition to two. Therefore, this particular command is of paramount importance during the execution of machining operations. The Mw07 code is the subsequent significant code. Additionally, it is employed to activate the refrigerant. However, it is advantageous to activate the refrigerant that was overlooked. Therefore, the M07 activates the failed coolant system, which is employed to apply a fine mist of coolant or lubricant during the machining process. Mw07 is frequently employed to chill and lubricate the cutting instrument and whatnot. Therefore, to initiate the flow of the missed coolant, simply enter the code mw07 into the program. The interpreter will activate the flow of the missed coolant whenever this code is executed. The M08, the subsequent critical M code, operates in a similar manner, activating the coolant exclusively. However, it will activate the fluid-type coolant. The fluid-type coolant will be activated. Therefore, the M08 activates the steam fluid coolant system, which then lubricates the machining area with the coolant or coolant mixture to dissipate heat and eliminate the tips. In contrast to the M07, the M08 is employed to facilitate a more substantial refrigerant discharge.

M08 - Coolant On (Flood Coolant)

Syntax:
M08

To employ this coolant, merely enter M08 into the program editor. So in the last two codes, M07 and M08, we encountered two critical terms: fluid type and mist type. And these are the two most frequently employed methods for chilling and lubricating the cutting instruments in this CNC machining. Mist coolant, also referred to as aerosol coolant, is a thin spray or mist of coolant liquid that is directed toward the cutting instrument. It is typically employed for this purpose. Milling and really mist coolant are examples of low- to medium-speed machining operations that are more precise and targeted, thereby reducing the quantity of coolant used and minimizing waste. It is appropriate for applications in which an inordinate amount of coolant could pose a problem, such as electronics manufacturing. The continuous passage of a substantial volume of coolant over the cutting instrument is the essence of fluid coolant. This can be accomplished by employing nozzles or other delivery systems. It is frequently employed in heavy-duty and high-speed machining operations, where it is employed to disperse heat and extend the life of the tool, such as grinding or turning. Coolant is capable of providing efficient chilling and chip removal; however, it may necessitate additional coolant consumption due to its lack of precision. This approach is optimal for applications that prioritize semiconductor evacuation and cooling. The selection of mist and fluid coolants is contingent upon the specifications of the machining application, material, and cutting conditions. Mist coolants are more cost-effective and environmentally friendly, whereas fluid coolants are necessary for heavy-duty operations and efficient heat dissipation. So, with time and practice, you will determine which type of coolant is more beneficial for machining operations, as per your needs. We will now proceed to discuss the M09, which is the subsequent significant code. To disable the refrigerant flow, these M09 are employed. The refrigerant system is deactivated using the M code M09. It may be a Mw07 or M08, and it disrupts the flow of coolant or mist coolant depending on the previous coolant command. Therefore, if you have utilized the M07 or M08 commands to activate the coolant, but wish to halt its flow, the M09 command is the most effective method for deactivating the coolant flow. In order to execute this command, simply enter M09 into the program editor. The interpreter will automatically recognize that it is time to stop the refrigerant supply whenever it encounters this command. I trust that you have gained an understanding of the significant value of these fundamental M codes in the efficient execution of machining operations. Some of the practical implementations of the G and M codes will be the subject of our upcoming modules. Additionally, we will explore the significant advanced G and M protocols.

# PRACTICAL EXERCISES ON G CODE AND M CODE PART 1

By engaging in specific practical exercises, we will gain insight into the application of G and M codes. To begin, we will work with the most basic component and execute the milling operation. For the sake of simplicity, we will disregard the radius compensation and operate exclusively in absolute coordinates. The CNC simulation will be employed to fabricate this component. Therefore, I recommend that you capture a screenshot of this image in order to enhance its clarity. Therefore, upon further discussion, you will have a more comprehensive understanding of the various dimensions and measurements that are available for this component. First and foremost, we must establish millimeters as the unit of measurement and install the appropriate machine for our endeavor. The settings option can be accessed from the primary menu. Click on it and then select "Settings." Therefore, you have selected the millimeter in this regard. To close the settings dialog, select "OK" as this tutorial is created in millimeters. Proceed to the file menu and select the "load machine" option to identify this particular machine. Additionally, you will receive a notification that the complimentary version exclusively includes milling center and returning machines.

Therefore, select "okay." Now, it is time to install the equipment. Select this milling folder.

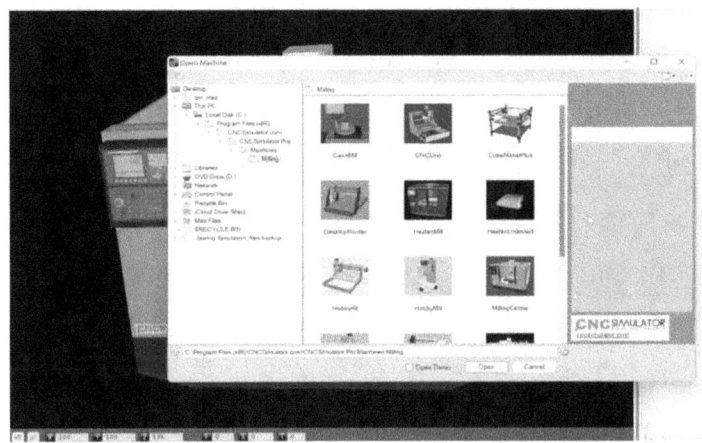

If the "open demo" option is selected in the dialog box that appears, kindly deselect it. Therefore, it is necessary to deselect these. Click on this milling center and then click on the "open" icon once more. Proceeding with the definition of our workpiece, we shall commence. The workpiece can be defined by pressing F2 and the keyboard or by navigating to these settings and selecting the Inventory Browser. Click on the Mill Workpieces tab located at the top of the dialog box. Here, you will observe that I have already incorporated a few of the milling workpieces. By employing these blue arrows, you can navigate through the various vertices that you have established. Click on the green plus icon to add a new workpiece if any of the other workpieces are unavailable.

I will select "add." The workpiece name should now be designated as "tutorial one." The subsequent step is to specify the dimensions of the x, y, and z axes. Therefore, we will input 75 for the x dimension, 70 for the y coordinate, and z for the ten. Okay, now mentally record the index number of the workpiece. In my instance, the index number of the workpiece. Tutorial one is forthcoming. It is; however, the value may differ in your specific situation. Therefore, you may observe the situation from this location. Click "OK" to dismiss the inventory browser and disregard the remaining configurations. Calling up our new workpiece from the program is now necessary. The D command dollar ad registry element is employed to accomplish this. Therefore, I will input the text into the editor. Therefore, we should initially indicate that it is our tutorial. Therefore, I will enclose it in parenthesis. Therefore, this is the CNC tutorial. It is now time to access the dollar advertisement for the workpiece category. The list of various options will be displayed as you enter "ed." Additionally, these alternatives may be navigated by employing the arrow keys. Therefore, our objective is to incorporate a workpiece into the registry. Therefore, we will select the "add registry type" option. Here, you will observe that the ad registry part command is accessible. Therefore, I will select it. Additionally, this pertains to the CNC simulator. At this time, accommodate an area. You will be presented with an inventory of various workpieces that you have incorporated into the registry as you allocate this space. In this instance, we intend to implement the tutorial option. Therefore, this is the object that is accessible at index number five. Therefore, I shall opt for it. Therefore, the interpreter will comprehend that the workpiece number will be utilized from the registry as a result of this line. And as we are aware, the machine zero is accessible at this particular crosshair symbol. So one is located here by the verdant hue. The workpiece will be positioned at this specific location by default. The portion will be added after the remaining dimensions are adjusted. Additionally, the command can be automatically inserted at the cursor in the editor by clicking on the inserted cursor icon. Additionally, we have previously addressed this matter during our session. We will now position our new workpiece 30mm away from the machine, with zero in both the x and y directions. To access our workpiece number five on x 30 and y 30, we write "dollar" and "register" from the machine zero, which is situated in the lower left corner of the machine table and is designated by a cross symbol. Comma 30, comma 30, part five. Therefore, it suggests that the workpiece accessible at the fifth index will be positioned 30 millimeters from the machine zero point in the x and y directions. Now, click on the "play" icon to observe your workpiece appear on the table. I will select on the "play" icon to accomplish this.

Therefore, you will be able to perceive this. This is machine zero. And the component is situated at a distance of 30 NM from the x and y axes. Starting from this mechanism, the zero position. Suppose that I remove these 3030 and subsequently click on the play icon again. Now, you will notice that this workpiece is situated at the zero position of the machine. Therefore, the position of your desired workpiece can be adjusted by supplying the x and y dimensions in the ad registry section. Therefore, I will once more input the number 3030. Click on the listen icon once more. I trust that you have a clear understanding of how to regulate the distance from this machine in the zero point. In the lower toolbar, select the "reset view" icon to focus in on the machine table. Therefore, I trust that you have acquired a more comprehensive understanding at this time. It is not possible to rotate the view by clicking on this simulation window with the left mouse button and dragging the cursor. The view will be panned when the right mouse button is clicked. Additionally, it is important to mention that the mouse wheel can be used to zoom in and out. So, let us commence with the CNC program itself. As we moved the workpiece in on the machine table and away from the machine zero, the lower left corner of the workpiece is now at x 30 y 30. That is not particularly practicable. Therefore, we should relocate the programming to 0.2 x 30, y 30, and z 0. This will lead to a zero position in the lower left corner of the workpiece.

By employing decoded z 92, we relocate the zero point of the programming process. Therefore, to relocate the zero point in the program editor, enter G 92 and subsequently specify x 30 by 30 and z zero. Next, press on the "play" icon once more and observe the zero point as it moves to the right of your workpiece. Therefore, I will select the "start simulation" option. Also, as you can see, the symbol with two cross marks is now visible. The first cross mark indicates the machine zero point, while the cross mark located here signifies the program 0.3. In the event that I wish to relocate this program zero to the x 30, y 30, and z ten, I would execute the following command: g 92 x 30, y 30, and then select the "play" icon to verify the program zero. Consequently, you can now see that the program zero is accessible in the upper left corner of the screen. What is it? The instrument is now required to mill the workpiece with the necessary control. Therefore, either enter F2 on the keyboard or navigate to the settings and select the inventory browser. Now, choose either my milling tools or embedded milling tools based on your specific needs. Therefore, we should initially evaluate the embedded machining instruments. The first tool that is available is the and mill tool, which has a mill diameter of six M and a length of 50. Therefore, the purpose of this instrument is to execute the machining operation. Therefore, we will implement this instrument. However, if any of the necessary tools are not present in these embedded milling tools, you can simply access these my milling tools. You can add the instrument to your requirements by tapping on the "add" icon. And we have already addressed these matters in our previous sessions. Therefore, I will opt for embedded machining equipment. Here, you will observe that the instrument has only one position.

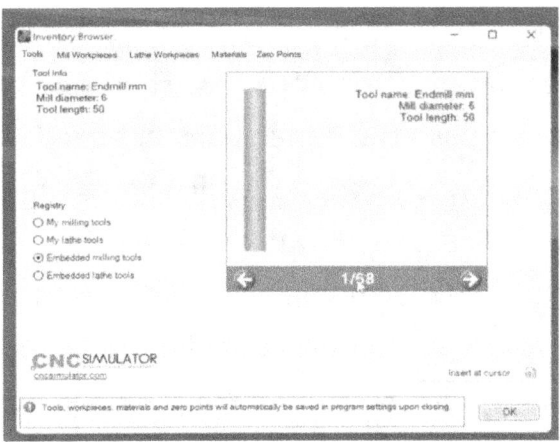

Additionally, this is the embedded utility. Therefore, to incorporate this utility, you may simple select the "Insert at cursor" option. Therefore, I shall opt for it. Therefore, you will observe that the 81 tool has been reintroduced to facilitate the modification of the tool. We are aware that the m code is M06 and that we wish to modify the instrument. That is the number 81. Therefore, it is possible to explicitly input D 81. Now, the instrument and the workpiece have been included. Therefore, we should now select the "play" icon. Zooming out will provide a more comprehensive perspective. Therefore, it is evident that the workpiece is also accessible in this location. The utility is also accessible. Therefore, we can commence the programming process to finalize the remaining portion of the pie. We will now examine the workpiece that we intend to machine. Therefore, these are once more the necessary dimensions. We should revisit these simulations once more. So, we have initially reset the program to zero. This is the time to establish the subsequent critical parameters. Afterward, we will commence the manufacturing process. Therefore, we should initially establish the dimensions to be in M. Additionally, you may utilize D 21, which limits the use to three millimeters. Press the Enter key. The subsequent step is to execute the machining exclusively on the x-y plane. Therefore, we will enter G17, which displays the plane selection for the x-y plane. Press the Enter key. Next. We are once again employing the absolute programming only society 90, which establishes the model to the absolute programming. Again, I will enter D G 94, which denotes the minute input. As we have established all of these parameters. In order to optimize the machining process, it is necessary to activate the coolant. To achieve this, we will implement the DM08 option. Click on the "OK" button. Initially, we intend to relocate the D tool to a location that is in close proximity to the workpiece. Therefore, the G00 will be employed for the swift movement. Additionally, we will designate the x, y, and z coordinates as zero, zero, and twenty, respectively. To obtain a more comprehensive understanding of this code, click on the "play" icon. Therefore, it is evident that the instrument is now more readily accessible in relation to these workpieces. The subsequent step is to activate the spindle. The m zero will be employed to code for this purpose. Additionally, the spindle speed is set to 2000, and the input rate is set to 0.02. Thus far, we have outlined all necessary parameters and initiated the spindle. In order to gain a more comprehensive understanding of the tool's movement, we will now navigate to this view icon. Additionally, the option feed will be activated at this location. Return to this program editor. Initially, our goal is to determine the appropriate outer contour of the millwork. Therefore, it is imperative that we investigate this matter thoroughly. What is the definition of flawless machining? Therefore, I will enter G0 one and the thickness of the workpiece, which is d ten. Therefore, we must transition from the zero portion of the program to the minus ten section. Therefore, I will enter g0 one z minus, if that is acceptable. Execute this code once more. Therefore, you will observe that the tool mode is set to 10 a.m. below to the zero position of the program. Once more, we will examine the geometry. Therefore, it is evident that the initial step is to machine the surface horizontally in the x direction by 60 meters. Therefore, I will input g0 one x 60. And as you can observe, the G01 comma d has already been implemented in this location. Therefore, it is straightforward to input the value of 60 as x. Therefore, the interpreter will automatically comprehend that it is necessary to employ the G01 command exclusively. The subsequent step is to machine the component with a clockwise rotation, which has a radius of 15 d. I will enter G02 for this purpose. For the end coordinate of the point and clockwise circular interpolation. In other words, x has a radius of 15 and a size of 75. Therefore, we should select the "play" icon. Therefore, you will notice that we have executed the machining process in a consistent manner. After that, it is necessary to advance 55 meters in the y direction. Therefore, as the code transitions from g zero 2 to 0 one.

It is necessary to reiterate the g zero block and specify the y coordinate as y 55. The subsequent step is to ensure that circular interpolation is conducted in the appropriate manner. I will enter G02 and set the x coordinate to 60, the y coordinate to 70, and the radius to 50. To gain a more comprehensive understanding of this section of the program, click on the "play" icon once more. The machining has been executed flawlessly up to this point. Once more, our objective is to advance in the direction of X. And as we are utilizing the absolute programming. Therefore, it is necessary to supply the x coordinate from the program's zero point. Therefore, I will enter G01X. It is necessary to perform machining in the incline line subsequent to this. The coordinates of x zero and d y 55 are indicated by the termination point of that inclined line. Therefore, I will enter x as zero and y as fifty. Then, machining is necessary up to y 15, and finally, machining is necessary in the inclined line, which is the terminus at x ten and y zero. We should conduct these simulations once more. Therefore, I will select "Play Simulation" and conduct a more thorough examination of the machining process. Please be advised that we have not taken into account any compensation for the entire radius. Therefore, the purpose of the machining is to elucidate the process by which the core is simulated. Therefore, let us once more examine the geometry. Therefore, it is evident that the initial incision was a straight line, followed by a counterclockwise rotation. Then, the rotation is performed in a clockwise direction. Once more, this is a linear cut in the y direction, followed by circular interpolation in the clockwise direction. And once more, a straight line movement in the x direction. The final step is the inclined cut, followed by another downward movement in the y direction and the final inclined cut. Therefore, we have already executed these numerous operations. Therefore, the geometry that is necessary has already been acquired. Therefore, we will proceed to the simulation location. As soon as our machining is finalized. Initially, we will deactivate the spindle, and I will enter the code M05. The coolant will be deactivated once more. Thus, I shall type M09. We will now swiftly remove the instrument from the workpiece. So, I will input G00. The initial step will be to elevate this instrument. I will input z 60 from this workpiece. Then, we will remove this instrument from the workpiece and from the x-y plane. Therefore, I will input x hundred and y hundred. Ultimately, the program concludes. Therefore, I will input M02. In order to acquire a more comprehensive understanding of this simulation, we will re-run the entire code. Therefore, I will select the "start simulation" button. Therefore, we will now examine the component that has been machined. We will employ the left and right mouse click to obtain a more comprehensive view of this machine. The component. Once more, we will relocate it downward. Enlarge the view. Allow us to rotate it slightly. As evidenced by the green lines, the fields are evident. This is the component that has been machined using this code. Therefore, we should examine the necessary geometry. Therefore, it is evident that we have generated comparable geometry by employing the CNC code provided. Exactly. We have not adhered to all of the coordinate points; rather, we have created the code to gain a more comprehensive understanding of how to utilize the various g and M commands to execute the machining operation. I trust that this tutorial will provide you with an understanding of the process of loading the workpiece and tool. Subsequently, you will be able to conduct all of the machining operations by composing the necessary G-code in the editor. In our subsequent session, we will continue to practice the next component.

# PRACTICAL EXERCISES ON G CODE AND M CODE PART 2

We will proceed with the practice of the subsequent component. Our current goal is to develop a controller that is similar to the one depicted on the screen. Therefore, you may obtain a more comprehensive understanding of this design by capturing a screenshot. While we are determining the various coordinates to be used to manufacture this component, let us revisit the simulation. Therefore, our primary objective is to equip the machine in order to execute the machining operation. In our situation, it pertains exclusively to the milling machining center. I have already loaded that computer during our previous session. We have already addressed the process of loading the necessary machine. Press F2 on your keyboard or navigate to these settings and select the inventory browser. Click on the mail workpiece to reveal that I have already included the five distinct workpieces in accordance with my specifications. Therefore, I will proceed to the fourth workpiece. These workpieces have dimensions of 100 d x, 80 y, and an acceptable z dimension. Therefore, in your requirement, you must select the ID icon, provide the workpiece name in accordance with your requirements, and specify all necessary dimensions.

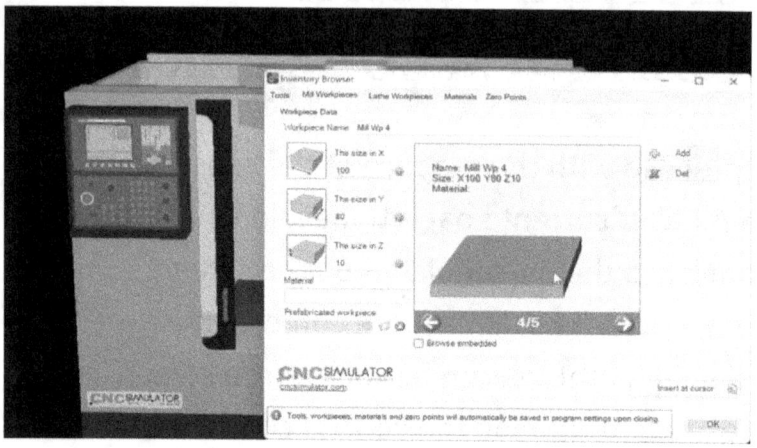

In my situation, I have already incorporated this workpiece; therefore, I will merely select the "insert at the cursor" option. Therefore, the component has been incorporated into the editor as a result of my selection. Therefore, it signifies the registry addition component. For the time being, I intend to relocate this component from machine zero. Therefore, it should be situated 30 nautical miles (NM) from the x-axis and 30 meters from the y-axis. To verify that the workpiece has been appropriately added, select on the "play" icon. We can either zoom in to obtain a more detailed view or select the "Reset View" option to obtain a more detailed view of the white piece section. Therefore, this location provides a more comprehensive perspective on the workpiece's displacement from the zero point of the machine. Subsequently, we must incorporate the instrument in accordance with our specifications. Therefore, I intend to utilize only the six Middletons with the same diameter. Therefore, in order to modify the tool, I will initially enter M06, as I am exclusively employing the embedded tool. Therefore, I will input it. The following is a comprehensive inventory of the various N mills that are currently available. Therefore, our goal is to exclusively employ six MB and the mill. Therefore, that is the initial alternative. This is intended to be a D flat and mill two. Therefore, I will select these by pressing the Enter key. Subsequently, it is necessary to establish D program zero in order to optimize programming. In order to establish the D zero offset, we will employ the D G 92 code. Additionally, we intend to establish the offset at x 30, y 30, and z 10. Therefore, this coordinate will serve as the program's zero. Click on the "play simulation" icon once more. And now, you will observe the distinction in the crosshairs symbol that is accessible. Therefore, this is the machine zero. And the program zero is indicated by these green crosshair symbols. Therefore, it is necessary to furnish all coordinates from the zero point of this program. Therefore, the absolute programming model will be implemented in this instance. Next, we will establish the remaining parameters. Thus, I will be selecting a g 17 in the x-y plane to serve as input for the d g 21 two machines. Subsequently, we will exclusively employ the absolute programming mode. Therefore, I will input G 90. Lastly, in order to establish the fields per minute, I will enter G 94, which is beneficial for evaluating the feed in accordance with the D feed per minute. Subsequently, we will initiate the flow equivalent prior to commencing machining. Therefore, I will enter M08. This will be followed by a rapid movement of the D tool toward the workpiece in order to initiate the machining process. Therefore, I will input G00 for the swift movement and specify the coordinates as x 0, y 0, and z 20. Once more, select the "play" icon. Therefore, upon the conclusion of this simulation, you will observe that. This instrument is now more readily accessible in proximity to this workpiece. Currently, we are interested in machining the entire outer aspect of the workpiece. Therefore, it is necessary to have a depth of ten meters. Therefore, we will commence machining at a depth of 10 a.m. initially. Therefore, in order to execute the machining process, it is necessary to activate the spindle. Therefore, in order to activate the spindle, I will enter the value of M03 spindle speed, which is one thousand. And if we require it to be 0.01 in addition to moving this straight line, we will select Z01 and set the z coordinate to t minus. In other words, it is accessible in the negative direction from these program zero points. Therefore, click on the "play" icon to determine whether the program has been effectively operating thus far. So far, we have not encountered any errors. So the program has been effectively implemented. Now, let us rerun the code and assume that I inadvertently entered -50. Therefore, the crimson line is used to emphasize this particular line. Similarly, the z value of -0.31 is outside the acceptable range at the bottom. Therefore, this is the appeal of employing this CNC simulator. The error will be immediately displayed if you have input incorrect coordinates or coordinates that are not feasible for this particular environment. Additionally, this form of error will facilitate the enhancement of our programming. Therefore, I will select "reset." Next, we will adjust the dimension of d to z minus ten. Click on the listen icon once more. Presently, what is in operation? We now have a more comprehensive understanding of the geometry that we are going to machine. Here, we will provide D coordinates. Initially, we intend to advance in the y direction. Therefore, I will input DG01, and the radius of the d cutter in the y direction is D3M. Therefore, it is imperative that we ensure the tool's correct movement. Therefore, the device consumed 20 megabytes. We will refer to d 23 M as a result of the milling cutter's radius. Therefore, I will enter y 23 in the space adjacent to the cut D position. We will be proceeding in the direction of X. The x-directional movement of the ten M is evident, as the tool has a diameter of d6. However, the radius of that particular circular means that it is inherently 33 x. Therefore, I will once more relocate the tool to the seventh coordinate. In the same manner, I will enter the y coordinate as 37 for y, and the tool will be relocated to the zero point at x. In order to ascertain the concept, we should attempt to execute the code. Whether or not the component is chopped. Therefore, it is evident that the necessary opening has been precisely made. We will now proceed with our subsequent machine. Presently, it is imperative that we relocate. Currently, it is necessary to remove six TM from this program 0.2. Therefore, I will input y=60. Next, we must generate a circular interpolation with a radius of 15 and end coordinates of d x 15 and y 75, which is to be applied clockwise to D. I will input G02X 15, y 75, and radius 50. Therefore, we should execute the code once more. So far, the D component of the computer has been functioning correctly. The subsequent component is to be divided up to 20 NM in the x coordinate. Therefore, I will input D01X 20.

The subsequent machining process necessitates a radius of 50 and terminal points with a diameter of 58. Therefore, it is being rotated in a clockwise direction. The coordinates x 50, y 80, and radius d will be selected, followed by Z02. We should execute the code once more. Therefore, we should adopt a more comprehensive perspective, as illustrated by this component. With this code, we have generated half of the geometry, as you can observe. We will now attempt to trim the remaining component in a clockwise direction once more. Circular interpolation is necessary, with the end coordinates being the x 80 and y 75. I will enter g=0, two x=80, y=75, and a radius of 50. The subsequent tool must be relocated to the 85 M position in the x direction. We should execute the code once more. Subsequently, as illustrated in the geometry, it is necessary to implement circular interpolation in a clockwise direction to generate the circular contour. Therefore, we enter G02X hundredy 60 and set the radius to 50. We will perform 20 additional straight line movements in accordance with the remaining coordinates of the components. I will input G01Y 37. The subsequent movement in the x direction will be 93. The y-axis should be 23 and the x-axis should be 100. And ultimately, we will return to the program 0.8 and set the y dimension to zero. That is the zero. We should once more align the mechanism with the component. Therefore, I shall initiate the simulation. Therefore, it is evident that we have returned to the initial starting point and have completed the necessary machining. The spindle will be halted at this time. Then, once more, cease the discharge of the coolant. Therefore, input M09. We will now move the tool away from the workpiece in the z direction and then again move it away from the workpiece in the xy plane. Therefore, I will input X00Y100. Additionally, upon the conclusion of the curriculum. So, in conclusion, we will enter M02. Finally, let us re-run this simulation and examine it more closely.

Therefore, you will acquire an understanding of the process by which various coordinates are machined in accordance with the instructions that we have established. We should examine this machined component more closely. In order to obtain a more comprehensive perspective, we should initially select the "Reset View" option. Once more, utilize the magnify function to enhance your perspective. Therefore, we will now compare this component to the geometry that we have employed. Therefore, as you can observe, we endeavored to decompose it in accordance with the dimensions specified in the geometry. Additionally, we have neglected to account for tool radius compensation in numerous locations. Therefore, we endeavored to achieve a comparable shape, which is outlined in the geometry. I trust that you will gain a more comprehensive understanding of the various G and M codes and their application in the machining process. Additionally, I hope that you will be able to insert the workpiece and D tool into the machining center in accordance with the necessary specifications. In our subsequent videos, we will explore the D advanced G and M codes, which are more advantageous for the creation of complex geometries or the execution of repetitive tasks.

# PRACTICAL EXERCISES ON G CODE AND M CODE PART 3

NCG and M core utilization. In this session, we will endeavor to generate the geometry that is depicted on the screen. For these, the workpiece is provided with a dynamic depth of d and a length and breadth of 83 m. Therefore, our primary goal is to develop the circular component with the ADM diameter. We will attempt to construct five distinct openings at specific locations after the circle has been formed. The four openings are provided at a diameter of 6 p.m., as you can see. and a central hole is included, with each hole having a diameter. Therefore, in order to proceed with the programming, we will navigate to the CNC Simulator Pro. As you can see on the screen for your convenience. The code has already been included in the editor, and we will now endeavor to comprehend its usage line by line. As you can see, the CNC header is initially provided. The milling center was employed to generate this code, and the unit is set to millimeters. In the same vein, the material dimensions are 83 by 83 by ten M. Therefore, our initial obligation is to equip the machine. We are aware that you have the ability to access this file. Additionally, the milling machine must be loaded from the load machine option. Next, the workpiece is added. The code has already been incorporated. The dollar add registry is that. The index of part seven is seven. In your specific situation, these indexes may differ, and the position of the workpiece from the machine is indicated by the subsequent 30, 30, and 30. None. And as you can see, the machine zero is present and denoted by these green clusters. The G 92 is the subsequent series. Here, D 92 is employed to establish the program zero for our workpiece. Therefore, we should initially verify that the white portion has been properly affixed. Therefore, I shall execute the code. Therefore, as illustrated here, the workpiece has been incorporated, and the end points of the workpiece are situated 30 NM from the x axis, 30 NM from the y axis, and ATM from the z axis. Therefore, it is accessible at a distance of 30 meters from the table. Assume that our objective is to preserve the distance from the table support. Given that we are endeavoring to form openings. Therefore, it is reasonable to presume that we have implemented specific supports to maintain this particular workpiece at this precise distance.

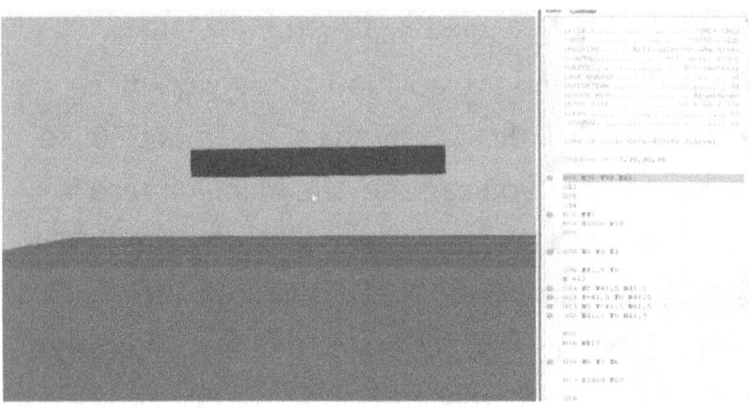

Currently, we are examining this particular workpiece. The subsequent step is to activate these millimeters for use. Therefore, the code g 21 has been coded. Next, the G90 code denotes absolute programming, while the G94 code provides the US steel input per minute after the workpiece has been loaded. The tool's subsequent objective is to be loaded. Initially, we will employ the instrument with a six-MB capacity to generate the circle from the rectangular component. Therefore, the term M06 81 has been employed in this context. Therefore, the embedded instrument is denoted by 80. And the initial instrument with a diameter of m that we have employed. Therefore, I will proceed with our simulation. Therefore, I will once more select this play icon. We have now arrived at the stage of selecting one of the two. The M04 is employed in the subsequent line to activate the spindle in an anti-clockwise direction. The spindle's speed is 1000, and the feeder supplied is 20. The subsequent M08 code signifies the activation of the coolant. Consequently, the tape appears as a single line on the screen; however, the water-like symbol is not visible. Therefore, we should resume our simulation. The tool is currently inserted into the machine, and the coolant supply is operational. This is evident in the symbol that resembles a drop. The utility is currently accessible. Initially, we will endeavor to swiftly relocate the tool to the center of this particular workpiece. And as you can see, our goal is to develop the circular component. Therefore, we endeavor to place the program zero at the core of this particular project. In order to preserve the "0" value of this program, we have implemented the following line: G 92 x 70, y 70, and z 40. Therefore, the center position is in the x coordinate from these machine zeros. It denotes that the y coordinate of the segment m is 70. Additionally, the 48 is determined by the table's base. Therefore, the distance specified is z 40. Therefore, it is possible to witness these green symbols once more, which serve as indicators of program zero. Therefore, we will furnish all coordinates from this particular center point. Therefore, in the subsequent line, we will be relocating this instrument closer to this particular center. Additionally, we will be repositioning the tool in the X direction to remove it from the center of the next line. Additionally, it indicates that the value is 41.5. Therefore, as you can see, we are employing the instrument with a six-inch diameter. Therefore, its center point corresponds to the coordinates specified in this document, as our goal is to generate a circle with a diameter of ATM. We are specifying a radius of 41.5, which corresponds to a diameter of 83 M. This is due to the fact that we are specifying the X direction and movement to 41.5.

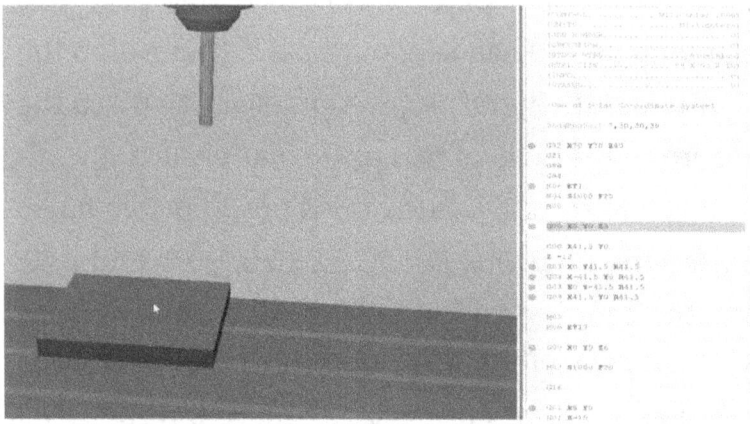

Therefore, the instrument automatically eliminates the three MB material from a specific side, as it has a three MB radius. Therefore, we should proceed with our simulation. Therefore, it is evident that the instrument initially arrived at the center position. Subsequently, it moved swiftly in the x direction by 41.5 M and in the z direction by D 12. The instrument is moving downward to the center position, as indicated by the value of -12. Our current goal is to generate a circular arc with a radius of 41.5 and an anticlockwise orientation. So, that is the reason we have composed the fundamental G0 three. The end coordinate will be zero, and the radius and d y will be 41.5. By default, we will select d to be 41.5. Therefore, we will attempt to execute this particular line. Therefore, we have eliminated this particular supplementary material by establishing this act. To proceed, we must once more relocate to the subsequent position. Therefore, the x-coordinate will be -41.5 in this instance. Additionally, the y-coordinate will be negative. The circle radius will remain consistent at 41.5 in all cases as we construct it. In the same manner, we have provided the coordinates for the third point and the final point, which is a repeated point. Therefore, we will execute the code in a sequential manner. Once more, we will construct the arc for the subsequent point. Finally, we return to the same point from which we began. Therefore, it is evident that we have generated this particular circle, which is defined by the ATM diameter of this rectangular component. The spindle stop is indicated by the M05, which is located after the quarter end. Lastly, the subsequent instrument for piercing purposes must be chosen. Therefore, we have implemented the DM 064. So once more, let us zoom out and attempt to relocate these specific components in order to more thoroughly examine the tool. Therefore, it is evident that the instrument has been modified. Additionally, let us examine this particular instrument in greater detail. Therefore, it is evident that the pointed tool is the one that indicates the piercing tool. In this 17th index, we have chosen the embedded drilling instrument.

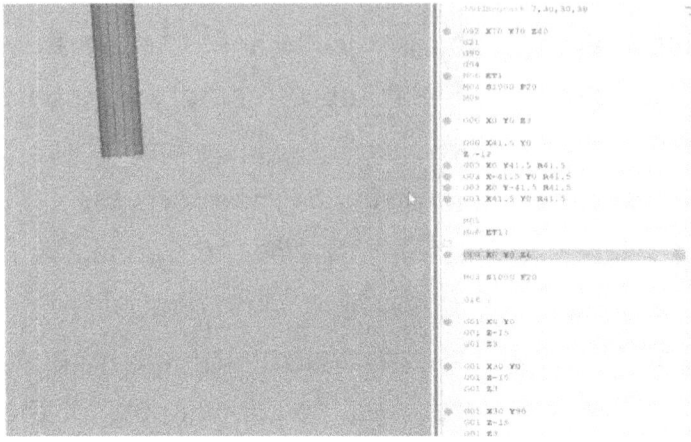

We have chosen this instrument due to its D denim diameter. The utility modification has now been finalized. Additionally, it is necessary to relocate this instrument closer to the center. The line X0Y0 and DZ6Z6 are the next lines that we have written. This indicates that the center position tool will remain at the same distance from this point forward. Next, we have activated the spindle and the subsequent critical chord that you will observe. The G 60 is the item in question. Therefore, let us revisit the geometry that we wish to establish. Consequently, we have initially developed this circular component with an ATM diameter. Now, as you can see from this central position, the location of all four openings is on these six barometer circles. Therefore, in our situation, it is more convenient to specify the center point of these openings and conclude the drilling process. Additionally, the radius and angle at which these various instruments are accessible can be employed to determine their location. Therefore, in order to capitalize on this particular combination, we will endeavor to implement the polar coordinate.

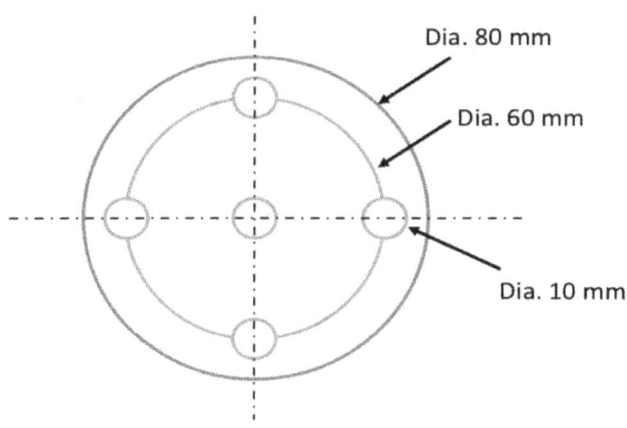

This will be extremely beneficial in the completion of this particular task, in which various necessary entities are accessible at a specific angular position. Therefore, we will initially activate the polar coordinate. Return to Resistance Simulator Pro. Therefore, the G 16 is the return character in the subsequent line. Additionally, this G 16 signifies that we are activating the polar coordinates. As we activated the polar coordinate. It is crucial to comprehend this. The radius is represented by the x dimension, while the angle is represented by the y dimension. Therefore, it is necessary to provide specific openings at the specific radius and angle. Therefore, we should proceed with our simulation. I would like to reset the view. Therefore, you can enjoy a more optimum perspective from this location. Currently, these tools are accessible at DX0Y0, and it is 6 a.m. away from this center. First, we will only create the opening at the center position. Therefore, we will relocate this particular instrument in the negative direction along the z-axis. Therefore, the value of d z is -15. This is where you can see that the workpiece's profundity is solely ten meters. This is due to the angular section of the drill instrument. It is preferable to advance the tool beyond the depth of the workpiece. Therefore, in order to ensure safety, we have specified d -15. Subsequently, we will endeavor to relocate the tool from the workpiece at a distance of three meters. Therefore, we have referenced DZ3 in the subsequent line. Therefore, we should endeavor to execute these particular lines. Therefore, the initial opening is formed, and the instrument is retracted to a distance of three. Moving forward, the opening must be drilled at a zero-degree angle and a radius of 30 ml. Therefore, the radius 30 is denoted by d x 30. The machining process will commence at these zero-degree angles, and the machine will interpret the right-hand side of d as the zero-degree angle. Therefore, the following line, which is x 30 and y 0, denotes the radius of 30 and the angle at which the instrument must be moved. The degree is nil. Again, we will be supplying the pit up to d - 15 m. Then, the instrument will be limited to D three, and we can proceed with the simulation. Next. Currently, the opening must be drilled at a 90-degree angle and with a radius of 30 meters. Therefore, the forthcoming line will return x 30 and y 90. All right. The cavity will be created in the z direction by rotating the tool in the negative z direction and then returning to d three. Therefore, we should execute this code once more. We will now create the cavity at a 180-degree angle in the same manner. In the same way, we will create the cavity at a 270-degree angle. We have established the geometry with a diameter of d atm, and we have created five apertures with a diameter of ten. These holes are located at a specific location. Therefore, we should revisit our geometry. Therefore, your component has undergone all necessary reductions and machining, as you can see. Therefore, we will conclude our operation by removing the instrument and ceasing the operation. Therefore, we will return to the CNC Simulator Pro next line that we have previously employed. That is the turn of the polar coordinates, denoted by D 15. Activating d Cartesian coordinates once more. The spindle rotation is halted by the subsequent M05. Next, we will remove this tool from the workpiece after the spindle rotation has been completed. Therefore, we will employ these rapid motion and moving D tools using an x-direction of 50 meters and an invite direction of 50 meters. Ultimately, we will terminate the operation and deem the program to be complete. We have successfully implemented DM zero; therefore, we should conclude the simulation. We will now examine the D component that we have developed. In order to enhance our perspective, we should attempt to pan mu and zoom in and out. Therefore, it is evident that the necessary geometry is generated by executing this particular code. Therefore, it is possible to witness the utilization of the polar coordinate in this context, as the dimension is located at a specific angular position and radius. This will be extremely beneficial in the completion of the D program. Therefore, in order to comprehend the entire operation with precision. We will now re-run the entire code and examine it in order to provide you with a more comprehensive understanding. We will first eliminate these boost points in order to execute the code in its entirety. We will initially activate the D feeds and traverse in order to provide you with a more accurate understanding of the tool's movement in a particular direction after removing those points. We should once more recalibrate the perspective. Zoom in to obtain a more detailed perspective. I will now execute this code by selecting the simulation icon. Now, if we examine this code and the editor line by line, we will observe that distinct operations are executed. The milling process has been finalized. Drilling will commence. Lastly, the tool is retracted from the workpiece. This is where you can observe the process of machining. Additionally, various intakes and transverse movements are offered. I trust that you have a clear understanding of how to utilize these polar coordinates. Also, how will these CSI programing simulations assist us in performing the machining prior to the actual machining of the materials? In our subsequent meeting, we will delve deeper into these topics and the M course.

# PRACTICAL EXERCISES ON G CODE AND M CODE PART 4

Next, we will proceed with the subsequent exercise to facilitate a practical comprehension of the G and M codes. In this session, we are provided with a rectangular plate that is 160 6MM in length and breadth, and 18 m in depth. Our goal is to establish the geometry that is depicted on the screen. Therefore, the initial step is to generate a circular component from the rectangular component. The diameter is 150 meters. This diameter is to be provided to a depth of ten meters. Eight is the result of the creation of these. Next, we must fabricate a circular component with a diameter of 70 M and a depth of 8 MM from the top. Following the creation of this fundamental shape, the subsequent step is to create the perforations. As you can observe, these holes are situated on the imaginary circle with a radius of 55 and are located at a specific angular position. Therefore, as we are aware, these openings have a radius of 8 a.m. m. Additionally, these are accessible at a distinct angle from the fixed radius. Therefore, the utilization of polar coordinates will prove to be more advantageous. To generate this geometry, we will first access the CNC simulator Pro. Therefore, the milling machining center was initially equipped. This equipment has been installed. The subsequent step is to incorporate the workpiece. Therefore, in order to incorporate the workpiece, I have already generated the necessary workpiece in the inventory browser. The line dollar add required has been incorporated, resulting in a height of 86 index. Therefore, I will specify six. I have now provided the x, y, and z axis distance, which denotes the distance from the machine zero to the left extremity of the workpiece. This command establishes the CNC machine in absolute mode, which means that all subsequent coordinates will be interpreted as absolute positions from the machine's reference point. The following command denotes the absolute programing. The subsequent command is g 21, which establishes the input units as millimeters. The transmission rate mode is established at two millimeters per minute using D 94. Subsequently, a G 92 signifies the establishment of a new program zero point for the x, y, and z coordinates, with x equal to 113, y equal to 113, and z equal to 38. Therefore, this will serve as our reference point, and we will provide the absolute coordinates for the subsequent machining process from this point onward, as we have already supplied the necessary workpiece. The subsequent requirement is to establish the required to. In this instance, it is necessary to establish this circular component. So, our initial step will be to choose the intermediate instrument. Additionally, this navigator is present in the specified utility. Therefore, this command instructs the machine to replace the instrument number one, which is an animal with a diameter of six. Additionally, it is the embedded utility that is accessible. Afterward, the spindle is initiated, and this command establishes the spindle rotation direction as clockwise, the spindle speed as 2000 rpm, and the input rate as 2.2 per minute. The coolant system is activated by the subsequent M0 eight command. Therefore, I will now select the "play" icon.

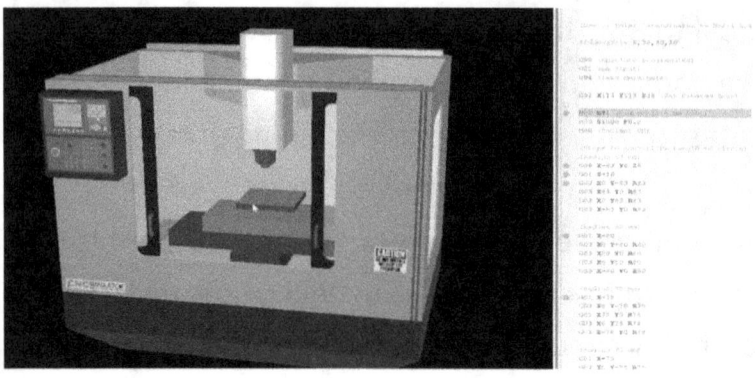

So, as you can observe, the workpiece has been incorporated. This is the zero location of the machine. Here, the program zero is denoted by the green crosshair symbol. Therefore, all coordinates will be provided from these zero zero positions. Additionally, the remaining machining will be completed. Therefore, the subsequent step is to establish the instrument after the workpiece has been established. So, as you can see, these M06s signify a tool transition. The animal with the 6MM negative is represented by this instrument. Therefore, we should resume the simulation. So, as you can see, the instrument is now also installed into the machine. Initially, we must convert this rectangle into the necessary circle for the continuous machining operation. Therefore, we have chosen the workpiece with a length of 166 meters. Therefore, we will commence with the machining of a radius of 80 meters, which is 83 meters (m). Therefore, the precise location of the tool's center point will be 83 meters. Additionally, because it has a radius of 3 millimeters, we will construct a circle with a radius of 83mm. Therefore, a point with a radius of 83 meters is indicated here. This represents the starting point of the manufacturing process. Therefore, we will initially relocate the tool from the center zero point to the outer point, with an x-direction movement of 8083. Initially, we have chosen the value of x to be -83. This implies that the machining process will commence from the left-hand side. Therefore, this suggests that there is a rapid pace of progression. Additionally, the tool will maintain a distance of six meters from the workpiece. We will proceed to conclude the linear interpolation at a depth of -18 m by employing the G0 one. Due to the fact that the entire workpiece has a total depth of 18, we will initially proceed by adjusting the diameter to a depth of ten meters. In order to accurately display the velocities and the initial hours, I will navigate to the view menu and determine whether it is the final hour. I will now proceed with the simulation. Therefore, as you can observe, the utility has relocated it to a distance of -83 meters. Additionally, it is situated six meters distant from this workpiece. We will now proceed with the machining process at a depth of 18 meters. And in order to fulfill our requirement to generate the circle, we will commence the machining process in an anticlockwise direction. Therefore, it is necessary to proceed in the negative y direction from this juncture, as the center point is situated at this location. Therefore, the y-coordinate of -

83 represents the negative y-direction. Therefore, we have specified a radius of 83 and a value of d y of -83. In the same manner, we will select all of the remaining four points to form the circle. Additionally, these points are indicated in this location.

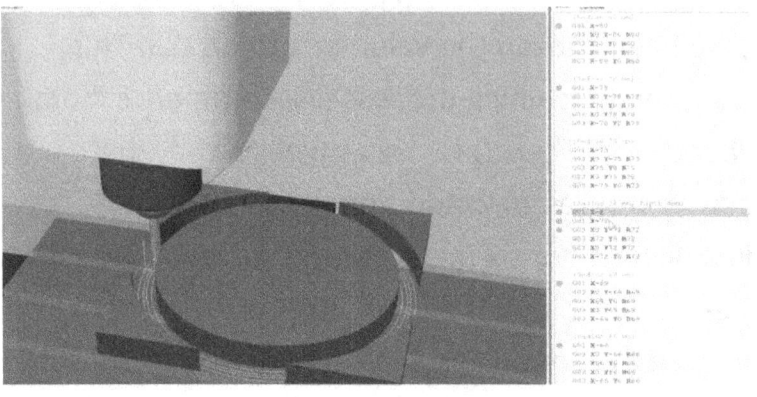

Therefore, we should proceed with the simulation. Therefore, the circular component has been generated and the machining has been completed up to the point of 83. And once more, I would like to emphasize that the tool's radius of three meters at the 83-meter site will result in a p-circle with an ATM diameter. Therefore, we should examine this circle. Therefore, it is evident that these circular components are generated. Allow us to recalibrate the view and pan, zoom in, and zoom out the component to obtain a more detailed view. As you can observe, the circular component is generated. The circular component must be constructed with a radius of 75 members. Therefore, we will persist with this procedure until the radius of 75 is reached. Therefore, we will decrease the radius by three NM in each operation and proceed to construct the circle. Therefore, this will yield the appropriate outcome and generate the property. Therefore, the subsequent circle will be constructed from the radius of the ATM. Therefore, we initially relocated the instrument to a distance of x -80 using G0 one. Additionally, we will execute the circular operation once more. Therefore, I will proceed with the simulation. Therefore, the radius will be diminished in this instance. In the same manner. We will now decrease the radius by three. Additionally, we will proceed with the machining process from the 78 MB point. Therefore, we should proceed with the simulation once more. So, as you can observe, we have generated a circular component. This procedure is intended solely for your convenience, in order to provide you with an understanding of the machining process. Subsequently, we must generate the circular component. Currently, the profundity is 8 mm from the top. To begin, we will modify the z distance to be d minus eight m. Also, we will proceed to generate the radius from 70 to m m point. Additionally, we desire to include a circle with a radius of 30 5MM at the summit. Therefore, we will continue the process of forming a circle until we reach the point. We generate the circular component with a radius of 35 mb for d. Therefore, we should proceed with the simulation. So, as you can see, the profundity has been reduced to minus eight MB. The subsequent instrument is relocated to a distance of -70 meters. Therefore, it is possible to observe the tool's movement. Again, we will commence the development of the circular component. Thus, we will commence the process of establishing the radius from these 72 m points. Therefore, we should proceed with the simulation. The machining is currently being conducted from the upper part, as you can see. And the profundity that is offered is solely 8MM. Then, relocate the tool to 60 9MM once more and create a D circular component to remove the material from the circular operation. We will now proceed with the operation at the D point of 66 M, and we will reduce the diameter from 66 to 63 again, to 6057. We will continue the operation until it reaches 38 M. Therefore, we are proceeding to the 38 MB mark, as the instrument has a radius of three M. Therefore, we will generate a circular segment with a diameter of 35 mm up to a distance of 30 8MM. Therefore, we will proceed with the simulation and examine this simulation in order to provide you with a more comprehensive understanding. Until now. The material has been removed from the outer section, as is evident. This final incision at the D point of 38 M will now serve as the circular component with a diameter of 35 mm.

Therefore, we should resume the simulation with this final edit. We have successfully constructed the lower section with the necessary diameter and the upper section with a diameter of 70 m, as you can see. Once more, return to the necessary geometry. Therefore, as you can observe, the upper portion of this object has a radius of 30.55 mm and a height of 8 mm from the center of the object. In the same vein, the height of this lower section is present. We are now required to create these openings, as the milling operation has been completed. First, we must modify the instrument in order to generate these openings. Afterward, we will employ polar coordinates. Therefore, once more, consult these professionals who specialize in NCC. Therefore, it is evident that the tool is being swiftly relocated from the workpiece at a distance of 20 meters from the summit. The spindle is then brought to the top and the M06 tool is employed to modify it. The drill tool, which has a D diameter of 16 M, was chosen as the anti-tool. Once

more, the spindle began to operate at a thousand revolutions per minute. Additionally, if it is necessary for it to serve as the focal point. Therefore, we should initially modify the instrument. Therefore, we will proceed with the simulation. Therefore, it is evident that the instrument has undergone a transformation, as evidenced by the modifications to its form. In the subsequent line, we will once more rapidly advance the multitool to the zero point of the D program, which is six megabytes from the center. To generate the gaps. We are aware that these openings are accessible at the same radius as D, but they are distinct. And we employ the polar coordinate. That will be the superior alternative. Therefore, we should proceed with the simulation. Therefore, you will observe that the instrument swiftly advances to the center position. It is six megabytes away in the z direction. The polar coordinate will be activated subsequent to this. The radius and angle will always be indicated by the x and y values, respectively, when the polar coordinate is activated. Therefore, let us revisit this geometry once more. Therefore, it is evident that these openings are accessible within a radius of 55.
Additionally, the initial opening is accessible at a 30 degree angle. The subsequent angle is a 90-degree angle. The subsequent one is also at a 150-degree angle. The subsequent opening is situated at a 210-degree angle. The subsequent hole is available at a 270-degree angle, and the final hole is available at a 330-degree angle.

Therefore, we will maintain the same radius, but we will modify the angle. Therefore, we should proceed with the code. Therefore, it is evident that the initial opening necessitates relocation. The radius of 55 meters. Additionally, at a 30° angle. And after moving to that location, the workpiece now has an overall thickness of 18 meters. Therefore, in order to ensure the safety of all parties involved, we will drill the shaft to a depth of 23. The instrument will be redirected back to the same distance after drilling the cavity again. Therefore, we should proceed with the simulation. The instrument will be relocated to a radius of 55 meters and entangled at a 30 degree angle. Therefore, we should proceed with the simulation. Therefore, the instrument is the mode. Drilling will be conducted, and it will be retracted once more. Therefore, G0 denotes a radius of 55 and an angle of 90 degrees. Additionally, the cavity is identical in terms of its profundity and diameter. Therefore, the value of z remains constant. Therefore, I will proceed with the simulation. Next, we will proceed to an angle of 150 degrees. Once more. 210 degrees is the next accessible opening. So, we will machine at an angle of 210. The subsequent opening is situated at a 270-degree angle. Therefore, we should proceed with the simulation once more. Additionally, the final hole is accessible at a 330-degree angle. Therefore, we should finish the final hole. The tool is now repositioned away from the workpiece after all necessary holes have been drilled. Finally, we will terminate the program, turn off the refrigerant, and halt the spindle. Therefore, we should conclude the simulation. We will now examine the workspace that has been established. Initially, we will deactivate the wheels and rapidly traverse in order to obtain a more comprehensive view of the component. Presently, examine this element. Therefore, it is evident that the initial two thicknesses are ten meters. The subsequent item is eight meters in length and has a diameter of 70 meters. The drill tool is employed to create all of the necessary openings at the appropriate angles. Therefore, I trust that you have comprehended the concept. What is the benefit of this EMC simulation in the creation of the component prior to the actual machining operation? Therefore, it is imperative to accurately visualize the entire machining process. We will begin by eliminating all the points. I will now activate the weights and rapid traverse once more. And now, once more, continue with this simulation and examine it to obtain a more comprehensive understanding of the various movements that are executed using the various G and M codes. Therefore, we should recalibrate the perspective. I will re-start the simulation after we access this component. Therefore, examine this simulation. Therefore, you will have a more comprehensive understanding. Right here. Machining at the lowest level. Completed. The machining process has commenced from the summit to a depth of eight meters. This is the final incision, and our objective is to reduce the diameter to 70. A component with a D radius of 35 M will be developed for this top. Subsequently, we will modify the instrument and execute the drilling operations. Consequently, the modification in the instrument is evident in this instance. Finally, the D polar coordinate is activated. We will initiate the machining process at a 30 degree angle. The final cavity has been generated, and the machining process will now commence. Therefore, in order to comprehend the use of this g and M core, we attempted to generate various components by employing these g n m algorithms. Therefore, it is possible to experiment with a variety of components in accordance with your needs. Also, develop this simulation prior to the actual machining process.

# ADVANCE G CODE AND M CODE

Therefore, we will commence our discourse with the more sophisticated G codes. Additionally, we will acquire knowledge regarding its syntax and framework. In the same vein, we will address the practical applications of the G and M codes. The G 41 is the initial G code that we will address. Additionally, it is advantageous for the compensation level of the trimmer. Therefore, the cutter radius compensation is activated to the left of the G tool by these G 41. Nevertheless, as illustrated in the accompanying figure, the cutter compensation is necessary when the tool is relocated to the left side of the component at the moment. We have observed in numerous examples that the center point of the tool will shift to the specified coordinate when the coordinates are provided. Consequently, the excess material is eliminated as a result of the tool's radius. Therefore, these g 41 offset the tool part by the tool's radius in the left direction, thereby compensating for the tools.

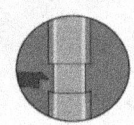 

### G41 - Cutter Compensation Left

- Activate cutter radius compensation to the left of the toolpath.
- It offsets the toolpath by the radius of the tool in the left direction, compensating for the tool's size.

Therefore, the g 41 command instructs this machine to modify the tool part to the left side of the program whenever it is executed. However, this command is typically employed when the cutting tool has a diameter and the machine must compensate for it in order to achieve the desired dimension. In fact, we are essentially extending the tool element outward by half of the tool's diameter to the left. In order to utilize this cutter compensation in a straightforward manner, it is necessary to input D 41. Subsequently, the CNC machine will automatically handle the compensation after the completion of any machining operation. In the same manner. The subsequent major command is the D 42. Also, it is beneficial for compensating cutters. Exactly. Therefore, the cutter radius compensation to the right of the tool is activated by these four. However, similar to DG 41 eight, the tool part is offset by the radius of the tool in the right direction, compensating for the tool's size. As illustrated in the figure, the compensation for the radius of the tool is necessary whenever the tool is moving to the right side of the required part. If compensation is not provided, it is evident that the coordinate must be accurately calculated by taking these radius into account. Additionally. Therefore, these compensations are exceedingly advantageous in intricate geometries. Therefore, the g 42 command is employed to modify the toolpath to align with the right side of the program. However, similar to G 41, it is employed when the cutting tool has a diameter and the machine must compensate for it in order to achieve the intended dimension. The toolpath is received outward by the alpha of the tool's diameter to the right. Thus, whenever you supply the cutter compensation machine, it is assumed that this tool portion is accessible at the center of the specific. Therefore, the commands used to modify it are G 41 and G 42. The dimension of the cutting instrument is taken into account by the tool part humans. When operating CNC machines, these commands are indispensable for attaining precise and precise cuts.

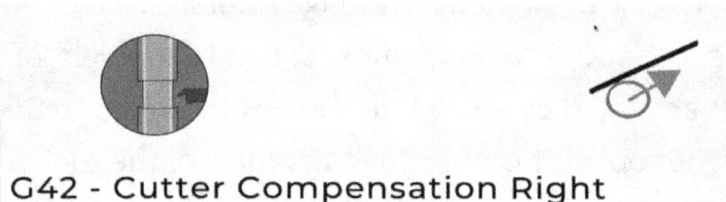

**G42 - Cutter Compensation Right**
• Activate cutter radius compensation to the right of the toolpath.

Therefore, in order to properly implement this cutter compensation, it is sufficient to implement this index T 42. Therefore, the instrument will automatically mitigate for the right-side movement whenever you enter G 42. As we previously discussed with the circular two, the same principle applies to the compensation of the tool snout. Therefore, these tool nasal compensations are categorized as left and right. Are these CNC machining commands employed to modify toolpath movements to accommodate the shape and size of the cutting tools? The CNC machine adjusts the toolpath to the left side of the program when the nose compensation is set to the left. However, this adjustment is contingent upon the shape and size of the tools.

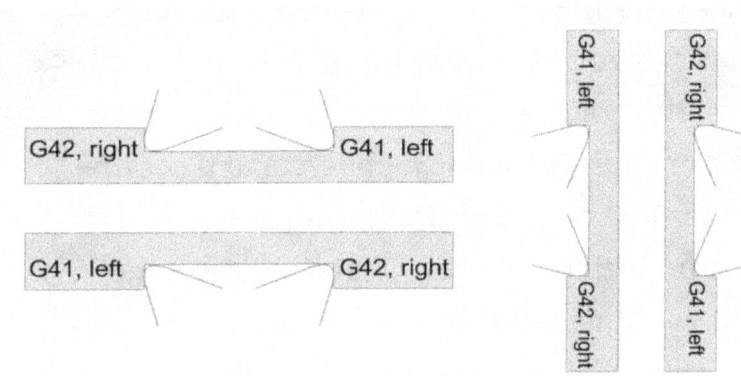

This command is frequently employed with tools that have intricate analysis shapes or are prepared to be accommodated in order to achieve the desired dimensions. The purpose of this command is to extend the toolpath by a specified amount, which is determined by the machine's controller based on the tools, noise, geometry, and the specified object. Similarly, tool nose compensation right is employed with tools that have a specific nose prepared or geometries that necessitate an offset to generate rate reduction. Therefore, this is a right-side command that extends the toolpath outward by a specified amount, as determined by the machine's controller based on the tool's noise, geometry, and specified offset. As illustrated in this figure, the necessity for left-side or right-side compensation is explicated through the figure. Therefore, the compensation for the left and right sides is contingent upon the tool's movement direction and side. Therefore, the movement is contingent upon the top, bottom, or possibly the left and right sides. It is necessary to select D G 41 and D 42. Therefore, the practice and utilization of these various codes will provide you with an understanding of the selection of this command. The subsequent critical command is located between 40 and indicates the cancellation of the cutter compensation. Therefore, this code is employed to negate the cutter compensation, which offsets the tool part by the tool's radius. It may be a G 41 or G 42, and it deactivates any previously applied cutter radius compensation. Consequently, the machine returns to the original programed tool part without any compensation whenever G 40 is executed. This is crucial when you wish to recommence standard toolpath cutting without any alternative. To employ this code, simply enter G 40. The CNC machine will automatically eliminate all previously specified offsets whenever this code is executed. Consequently, in summation, the G 41 and G 42 cutter compensation commands are indispensable for adjusting the toolpath to accommodate the tool's dimensions, thereby guaranteeing precision machining. When the compensation is no longer required, the cutter compensation cancel is employed to revert to the original uncompensated toolpath. Especially when working with tools that have specific dimensions or radii, these commands are essential for obtaining accurate and consistent results in CNC machining. We will address D and other G codes in our subsequent session.

# ADVANCE G CODES PART 1

Our discourse will commence with the subsequent G code, which is G 43. Therefore, this G 43 represents the entire length of the set in the positive direction. Therefore, the total length of the set is employed to account for the variations in the length of the various cutting instruments employed in the D, C, and C machines. Therefore, it is common for multiple tools to be used for machining purposes, with each tool having a distinct length and two of them having the same code and depth. The difference in length between these two length offsets may be the result of variations in a tool's manufacturing process or the necessity of substituting tools during a machining operation. The desired cutting depth and precision can be maintained during tool changes by modifying the tool and the length of the set. Therefore, in order to comprehend the length of the set in this instrument, it is necessary to be acquainted with various components, the first of which is the entire length of the set register. Therefore, this is a memory location within the controller of these CNC machines that contains information regarding the length of each tool, as well as a unique offset value specific to each tool. The next term is "offset," which serves as a reference point on the workpiece of one set. The origin of the part and the tool length offset are applied relative to the work opposite. The final term is "tool set," which is a device that is employed to determine the precise length of the tool. Before commencing a machining operation, it is customary to measure and establish the tool length. Therefore, the operator or the machine must update the tool length offset values whenever a tool change occurs in a CNC machining operation to guarantee that the new tool cuts at the appropriate depth. And the new tool is selected and loaded into the spindle of the first operator or CNC machine. The tool setter is employed to determine the precise length of the new tool. The measured length is then compared to the stored length offset for the tool in the tool length offset register. Therefore, we should examine the device. This is our standard instrument, which is already stored in the machine, as you can see. As you can observe, the length of the tool changes when it is added. In order to accommodate this change and maintain the same depth of cut with both tools, a tool length offset is necessary. The total length of the set for that instrument is now compared to the measured total length.

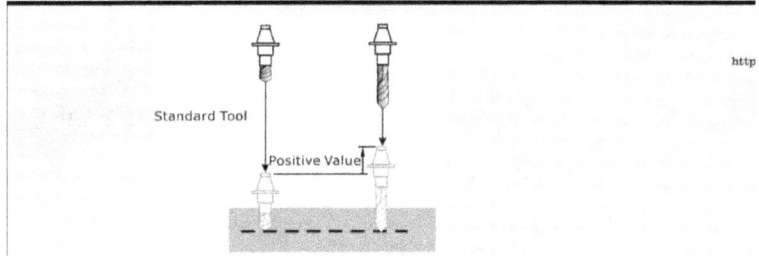

## G43 - Tool Length Offset Positive

• Used to apply a positive tool length offset.

In fact, the machining program is adjusted by utilizing the discrepancy between the measured length and the stored offset to achieve length offset. These modifications guarantee that the instrument operates in accordance with the intended path in relation to the work offset. The G course in C programs implements these two length offsets, and the most frequently used commands are D, G 43, and G 44.

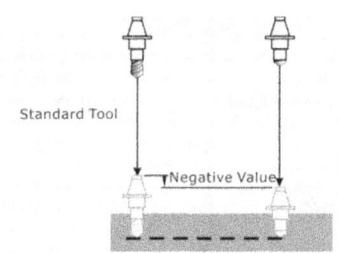

## G44 - Tool Length Offset Negative

• Syntax:
G44 H__ Z__
(H-code specifies the tool length offset number)

As you can see, the G 43 is employed to apply a positive two-length offset, which effectively reduces the cutting depth by bringing the tool closer to the workpiece. It is employed to modify the tool's position by adding a positive value to the tool, thereby enabling the tool to be positioned precisely. Therefore, in order to account for these two lengths, we are incorporating the positive value into this instrument, as evidenced by this reference point. Therefore, the g 43 command is typically followed by a h number. Additionally, the z-value. The offset is denoted by the z value, while the tool number for which the compensation is to be applied is specified by the h number. For instance, if we have entered z 43 H01Z5, the machine will be instructed to implement tool length compensation for tool one with an offset value of five. Additionally, the offset value is positive due to the use of g 43. G 44, similar to g 43, is employed to account for the variations in tool length that result from changes in the tool or the tool itself. Nevertheless, we are able to modify the cutting depth by relocating the tool away from the workpiece using g 44. It is employed to modify the tool's position by subtracting a value from the tool, thereby enabling the tool to be positioned with greater precision. These h numbers specify the tool number for which the compensation is applied, as the D 44 command is typically followed by a h number and z value. For instance, if the code is written as g 44, h two, and z two, the machine is instructed to apply negative length compensation to tool two, with the value being d. These g 43 and g 44 commands are typically used in conjunction with gate to land offset. In order to utilize these commands, the correct tool length offset values must be established in these CNC machines. The control system G 43 and G 44 commands are components of the standard G-code language that is employed in CNC programming. The CNC machines' control software and configuration may result in minor variations in specific syntax and functionality. In operations that involve multiple tool changes or lengthy machining cycles, where the location of the tool can affect the results, these two length compensations are essential for maintaining machining precision and accuracy. In conclusion, the G 43 and G 40 4RG codes are employed in CNC machining to implement total length compensation in a positive and negative manner, respectively. The G 45 code is the next significant code, and it denotes the positive tool radius of Z. So, the tool radius of Z is another fundamental concept in CNC machining that, similar to a tool length offset, is instrumental in ensuring the precision and accuracy of the machining operation. The CNC machine is able to maintain the intended machining dimensions by compensating for variations in tool radius, such as tool VR and various tool geometries, using the tool radius offset. For D two, as we previously discussed, the length offset is.

In the same way. The tool radius, offset register work offset, and tool center are the critical components of this CNC machine. These D 45 are the G-codes that are employed to apply a positive tool radius offset. The tool position is adjusted by adding a positive value to the tool radius, which enables the tool to be positioned precisely. This functions in a manner similar to that of D two. Land offset is applicable when the instrument has a diameter difference of D. G 45 is followed by a h number in order to utilize this code. And the tool radius of the set number is specified by these h numbers. Therefore, if we have written g 45 H02, it would suggest that the tool radius offset is positive. It is necessary to apply it to the instrument number. The negative tool radius offset is applied by subtracting a value from the tool radius to modify the tool position, as is the case with g 46, the g code used for precise tool machining. To utilize these, simply enter the number G 46, followed by the H number, which denotes the tool radius offset number.

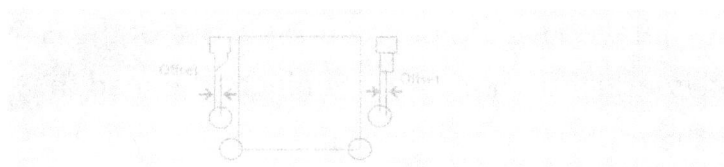

## G45 - Tool Radius Offset Positive

• Syntax:
G45 H__ (H-code specifies the tool radius offset number)

Therefore, the tool radius offset negative h to be applied to the tool number is indicated by the notation g 46 H01. One subsequent instrument of significance is d g 49. It is employed to deactivate the tool radius offset or the tool length compensation one that was previously activated using the D, G, 43, 44, 45, or 46 commands. These D 49 are G-codes that instruct the CNC machine to revoke any tool length compensation that was previously enabled with the G 43 or G 44 command. It essentially instructs the machine to cease the application of tool length offsets for the current tool. However, the primary objective of the G 49 command is to guarantee that no tool or compensation is in operation during the specific phase of the CNC machining program. This is especially advantageous when you wish to execute operations that do not necessitate the machine to modify the tool's position in response to two land objects. For instance, when performing non-cutting operations, moving between various workpieces, or repeating traversing, it may be advantageous to implement D 49. Use of the G 49 command is straightforward. To terminate the read the length or the tool radius compensation, you merely insert it into this CNC program. Therefore, its syntax is limited. G 49 must be entered. Therefore, the tool length compensation and the tool radius adjustments will be automatically canceled by the C machine interpreter whenever this line is executed. I trust that you have grasped the concept of the positive and negative length offsets, as well as the positive and negative radius offsets of the tool. These offsets can be cancelled using the D 49 command.

## ADVANCE G CODES PART 2

Our subsequent conversation will commence with Dee Dee, 73, who is referred to as Dee Dee 73. Command. Is the G-code employed in CNC machining to initiate high-speed drilling with crystal breaking? It is engineered to efficiently pierce deep cavities by swiftly retracting the tool to prevent chips from forming during each peak cycle. It automates the drilling process, which includes the high-speed retracting of the tool and the chip-breaking prongs. Following each bolt. The G 73 command is highly effective for rapidly excavating the defaults. In a deep, hollow operation, where deep evacuation can be difficult, it is used to optimize the drilling process by conducting rapid retracts and distorted weld periods between the pegs. This command assists in the prevention of chip accumulation and blockage, thereby enhancing the accuracy of deep drilling and reducing tool wear. We are subsequently employing the G 73 command. Parameters that specify the drill feed rate, retract plan, and Peg depth are typically included. Therefore, it is evident that these intakes are available for the G 73. So, you initially specify d g 73. The x, y, and z coordinates are then provided to indicate the position of the openings.

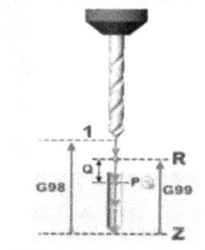

### G73 - High-Speed Peck Drilling Cycle
• Syntax:
G73 X__ Y__ Z__ R__ Q__     G73 X00 Y00 Z-50 R5 Q1 F100
X, Y, Z specify the hole's position.
R__ specifies the retract plane.
Q__ specifies the chip-breaking depth.

The retract plan is then specified, and the 3D deceleration depth is specified. Therefore, as illustrated in the figure, let us assume that we wish to drill a cavity at this precise location. Afterward, we will furnish the x, y, and z coordinates for this location. As illustrated in this figure, let us assume that we wish to establish a cavity at this precise location. So, initially, we furnish the x and y coordinates. The z position is then provided based on the required depth. The opening must be retracted after drilling. The instrument will return in order to achieve the desired position. That is achieved by specifying the retract plane. These Q ranges are primarily intended to maintain depth. Therefore, let us provide an illustration. As illustrated here, let us assume that we have a written record with the following values: D 70 3X0, y zero, z -50, r five, q one, and f 100. Therefore, the drill's position is represented by x zero and y zero. The initial z axis position for the drill is denoted by z -50, while the retract plane is defined by r five, which specifies the distance the drill should retract above the pit bottom. The number of peg cycles is denoted by q one, with one indicating a single pass, and the input rate for drilling is set at 100. Therefore, the xy machine will execute the drilling operation by repeatedly positioning itself at the initial opening location whenever it encounters the g 73 command. That is, the drilling will commence at Z-50, with the coordinates 0Y0. The tool will retract during each rest cycle. Preventing the accumulation of shards by inspecting them. The down field will be regulated by the specified bit rate during each peg, thereby guaranteeing the efficient removal of material. It is crucial to recognize that the CNC machine and the machining requirements can influence the specific parameters in the G 73 command. It is recommended that you consult the machine's documentation and modify the command accordingly. G 73 is merely one of the numerous drilling and hole-making cycles that are accessible in CNC machining. In conclusion, the G 73 command is a decoder that is employed in CNC machining for high-speed pig drilling with chip fracturing. It is particularly useful in deep hole drilling. It is essential for the accurate construction of deep cavities in the G 76 and D G 76 commands, as well as for the maintenance of efficient chip evacuation and the reduction of tool VR. The G-code "command" is employed in CNC machining to initiate a threading cycle. Threading is a frequent process in the production of parts that necessitate screws, bolts, or any threaded components, as it is employed to create threads on the workpiece either externally or interior. The G 676 command is primarily intended to orchestrate the threading cycle. It enables the specification of a variety of parameters, including the depth of cut, spindle speed, input rate, and thread pitch, thereby simplifying the process of generating precise threads. These a g 76 are indispensable for the development of threaded features. Several parameters are specified to regulate the threading operation when employing the D 76 command. Therefore, the fundamental syntax has been summarized here. Therefore, we will commence with d g 76. Then, we supply the x and z coordinates. These threading operations are executed on the turning machine.

Therefore, x and z coordinates are necessary. In order to utilize the g 76 command, it is necessary to specify a number of parameters that regulate the threading operation. Therefore, in order to employ this G 76 as a syntax, we must first enter d g 76.

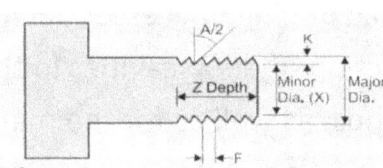

### G76 - Threading Cycle

```
Syntax:
G76 X__ Z__ P__ Q__ D__ F__ (or other relevant parameters)
G76 X0 Z-20 P1 Q0.25 D0.1 F0.05
P__ specifies the number of thread passes and types of thread.
X__ and Z__ define the thread's start point.
Q__ specifies the thread pitch
F__ specifies the feed rate.
```

After that, we specify the x and z coordinates of the beginning position. The p will be specified next, which specifies the type and the number of thread iterations. Consequently, Q specifies the thread pitch, d, followed by the number of threads, specifies the profundity of each thread, and f establishes the input rate for the threading operation. Therefore, I have included the following example: d 76 X0Z -£0.20. You identified 25 d point one and f 0.05. Thus, x zero denotes the initial x axis position, while z -20 denotes the initial z axis position, which is the depth at which the threading process will commence.

Subsequently, p one denotes the quantity of thread passes that would result in a single threaded position. Pointing to 25 establishes the thread's frequency. The feed rate for the threading operation is determined by the value of f 0.05, while the depth of each thread pass is specified by D point one. Therefore, the CNC machine will initiate the threading operation by positioning the initial thread start point at x zero and zero -20 when it encounters the TX 76 command. Subsequently, it will execute the threading operation. The feed rate remains at 0.05. The spacing of the thread reader is determined by the pit index 25. According to F 2.1, the number of passes and the profundity of incision for each pass are maintained. In order to comprehend the actual application of these 76 commands, we will consult a single sample video. So, as you can observe, the turning machine center has been installed. Additionally, the command will be visible on the left-hand side. Therefore, you will notice that we have implemented DG 76 in this instance. Additionally, the parameters for the initial block have been established through the implementation of DG 76. Therefore, the thread behavior is regulated by the three values that comprise the p parameter. The P010060 was employed in the example. Therefore, in order to comprehend this, we will disassemble the numbers. Therefore, the initial zero one represents the quantity of spring incisions. This implies that the machine can be programmed to accept a number of additional cards at the same depth to flatten the final thread after the thread card has been used. The runout angle is denoted by the subsequent zero zero, which is the angle at which the thread is released. Next, we have included these 60 degrees, which represent the inferred angle that is employed when the thread is entered. The profundity of the normal incision is denoted by q, which is expressed in hundredths. So the q 500 above represents 0.5, and r denotes the depth of the final or finishing cut, in the same manner as the second block is presented here. The final value of the x-axis is denoted by the X value. The 88 represents the final value of the z-axis. The block contains multiple parameters, including P, Q, and f. P denotes the thread depth, Q denotes the depth of the first out, Q denotes the depth of the first cut, and f denotes the thread. It is not necessary to extract each parameter at each iteration. These are some of the critical parameters for this d 76. The q parameter is the first parameter, followed by the x and z values of the second block and the f parameter of the second block. In order to gain a more comprehensive understanding of the machining process, we will commence the video. Therefore, examine this display. Therefore, the initial rotation is executed. Currently, the outer phase is being turned. As is evident. So, let us enlarge the image. And as you can observe, the strands are formed at a 60-degree angle. Therefore, I trust that you have gained an understanding of the manner in which the G 76 command executes this threading operation. The other G codes will be the subject of our next session.

# ADVANCE G CODES PART 3

We will commence our discussion with the subsequent G code, G 81, which denotes the drilling and cycle. Therefore, the G 81 command is employed in CNC machining to execute the drilling or basic hole-making operation. It is one of the most fundamental G codes and is employed to ensure the precise piercing of holes in the workpiece. The primary objective of the G 81 command is to automate the drilling procedure. It automates the drilling process, which includes the rapid positioning of the tool at the specified plant by piercing and retracting it. When employing the G 81 command, this G 81 streamlines and standardizes the drilling process. To operate the D 81, you must initially execute the Gat one command and subsequently specify the host position using the x, y, and z coordinates. This is necessary because you must specify a number of parameters to regulate the drilling process. Next, the correct plan was specified by specifying R, and the fit rate was again specified by specifying f. Therefore, let us provide an illustration. The 81 x ten, y 20, z -30, r five, and f 100 have been employed, as evidenced by the screen. So, whenever these machines encounter the Gat one command first, they will promptly move the mordi tool to the initial cavity location, which is the x ten and y 20. Then, it commences the drilling operation at z -30. Here, the feed rate regulates the downward movement of the drill as it penetrates the workpiece. It is important to note that the specific parameters in the g 81 command may differ depending on the machine entity link requirement. Therefore, it is imperative that you consult the documentation of the machine and modify the command accordingly. The G 81 command is a fundamental command that serves as the foundation for more intricate drilling and hole-making cycles, such as default drilling, tapping, and peg drilling. The quality and accuracy of the drilled hole are contingent upon D factors, including the selection of the appropriate drill bit, spindle speed, and feed rate. In conclusion, the G 81 command is a G-code that is employed in CNC machining for the purpose of conducting fundamental drilling operations. It automates the process of creating perforations in a workpiece by defining parameters such as the position of the t-tool, the depth of the hole, the input rate, and the location of the hole. The G 98 command is the next critical command that is employed during the can cycle to facilitate the return to the initial level.

## G81 - Drilling Cycle

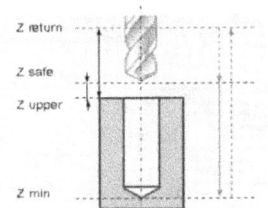

• Syntax:
G81 X__ Y__ Z__ R__ F__ (or other relevant parameters)
Ex: G81 X10 Y20 Z-30 R5 F100
X, Y, Z specify the hole's position.
R__ specifies the retract plane.
F__ specifies the feed rate.

The G-code G 98 is employed to return to the initial level following the completion of the drilling cycle. It ensures that the tool retracts to the initial Z level before proceeding to the subsequent operation. Therefore, as illustrated in the accompanying figure, our objective is to bore the cavity to a specific Z depth. This is now regarded as the initial level. Therefore, our objective is to relocate this drill bit from this point one to the necessary depth. The initial drill tool must be removed from the cavity and positioned at this precise location after the machining process is complete in order to produce the next hole. Additionally, the location must be such that it does not cause any damage. This g 98 command is beneficial for relocating the tool from the workpiece at a specific initial level from this final position. The primary objective of the DG 98 command is to establish a reference point for the commencement of the gained cycle. This reference point is essential for the preservation of the accuracy and repeatability of the machining process. These G 98 commands are essential in the Kan cycle, as they contribute to the preservation of consistency and accuracy in the machining process. The machine initiates the cycle from the precise and dependable position by returning to the initial point.

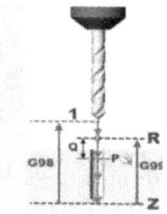

### G98 - Return to Initial Level

- Used to return to the initial level after drilling cycles (e.g., G81, G82, G83) are completed.
- It ensures that the tool retracts to the initial Z level before moving to the next operation.

The machine configuration is frequently established at the beginning of the can cycle, but it can be adjusted as necessary, as the specific initial point can fluctuate depending on the cycle definition. The can cycle can facilitate the programming of common machining operations, thereby reducing the risk of errors and saving time. The DG 98 command guarantees that the machine is always aware of its return location prior to initiating the cycle. In conclusion, the G 98 command is a G-code that is utilized in CNC machining to designate the initial point of a cycle. This is essential for the preservation of accuracy and consistency in machining operations that involve a predetermined sequence of G codes. Additionally, this code is to be implemented. You need only enter the command "G 98," and the interpreter will determine that the tool must be relocated to the 80th shell level after each operation. In the same vein, the subsequent critical command is d G 99, which signifies the return to the D reference plane. The G 99 command is employed to denote a cycle and to specify the return to the reference point or the tool change position. It guarantees that the tool retracts to the designated aircraft after the subsequent cycle is completed before proceeding to the subsequent operation. Therefore, it is evident that this one represents the initial position. However, if you prefer not to return to the initial position, you can establish a D reference plane to which the tool returns following the H drilling operation. And these are planes that are retractable. So once more, these planes should be kept away from the workpiece. It is crucial to maintain accuracy and repeatability in machining operations that the tool returns to a known reference point after completing the h guide cycle, which is the primary objective of the G 99 command. Upon completion of the G 99 command, the CNC machine will revert to the designated reference point. G 99 is crucial for the preservation of machining accuracy and consistency, as it ensures that the machine's tool is returned to a known reference point for the subsequent operation, as is the case with G 98. The precise reference point may differ based on the machining operation's requirements and the machine's configuration. Utilization of these types of cycles. The risk of errors is reduced and time is saved by the simplified programming of common machining operations. The G 99 command guarantees that the machine always knows where to return after concluding a task. Therefore, to employ this command, one must simply program G 99. Additionally, the CNC machine reiterates the tool to the reference whenever it encounters this command. Therefore, these supplementary G codes are essential for CNC machining programming, as they enable the flexible management of the coordinate system, input rates, and tool. However, these attributes render them indispensable for the attainment of precise and efficient CNC machining outcomes. Therefore, the Multiplicand cycle is employed in conjunction with these G 98 and G 99 commands. The G 81, the first significant Candu cycle, has been the subject of our discussion thus far, and it represents the straightforward drilling cycle. During our subsequent session, we will address additional significant manufactured cycles.

# ADVANCE G CODES PART 4

We will commence our discourse with the subsequent significant code, G 82. Additionally, it is the drilling cycle with T dwell. Therefore, the G 82 command in the C and C machine is utilized to specify the drilling with the dwell duration at the bottom of T, or it is a type of cycle that is employed to automate the entire manufacturing process with great precision. These data are comparable to G 81, but they include a 12-second period at the bottom of the board before retracting. Therefore, this is the sole modification that exists between these G 81 and 82. From a syntax and usage perspective, what are they? It is beneficial for applications that necessitate a brief halt at the bottom, such as the use of refrigerant or the performance of a cost-effective evacuation. Therefore, these dwell times facilitate chip clearance, lubrication, or other essential procedures, which are advantageous throughout the entire manufacturing process. When employing the G 82 command. In order to regulate the drilling cycle 12, you specify numerous parameters. Therefore, as demonstrated in these intakes, the host position is initially specified using the x, y, and z coordinates through the use of the G 82 command. The reference plane or retract plane is indicated below. The feed rate is denoted by F, while the dwell time is denoted by P in milliseconds. Therefore, as illustrated in this figure, let us assume that we wish to drill a shaft from this precise location and to a specific depth. Afterward, we will employ the g 81 command. However, this G 82 command is advantageous when you wish to maintain the tool at the bottom of the opening for a specified duration, as it is predicated on the use of D, G 98, or99. The instrument can be redirected to the initial position or the reference plane.

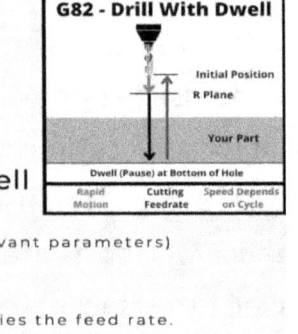

G82 - Drilling Cycle with Dwell
• Syntax:
G82 X__ Y__ Z__ R__ F__ P__ (or other relevant parameters)
EX: G82 X10 Y20 Z-30 R5 P500 F100
X, Y, Z specify the hole's position.
R__ specifies the retract plane, F__ specifies the feed rate.
P__ specifies the dwell time in milliseconds.

Therefore, in order to comprehend this, we will provide an example. As evidenced by the following, we have specified 82 extend y 20 Z -30 R5 E 500 and F 100. So, when the D 82 command is encountered by this machine, it will execute the operation and swiftly position the tool to the initial cavity location at x ten and y 20. After that, the drilling operation is initiated at d z -30. D downward movement is regulated by the feed rate f 100 as the drill penetrates the workpiece. After the drill has reached the specified depth, it will dwell or pause for the specified duration at the bottom of D. Subsequently, it will retract back to the reference position. The tool will be retracted from the hole after the dwell period, which is 500 milliseconds. The retraction position is contingent upon the use of the G 98 or G 99 command. In conclusion, the D 82 command is a G-code that is employed in CNC machining to drill with a specified dwell duration at the bottom of the workpiece. This method enables the automation of entire manufacturing processes while also incorporating controlled interruptions to address specific process-related requirements. The GA 83 command is the next critical command in CNC machining, and it is used to specify a peak drilling cycle for drilling deep cavities. Fake drilling is a technique that is particularly beneficial for deep hole drilling, as it entails drilling deep foils with multiple passes or is intended to remove chips from the cavity. The G 83 command is primarily designed to automate the peak drilling procedure. It enables the specification of parameters, including the back depth, feed rate, and dwell duration at the bottom of the opening, thereby reducing the risk of chip entanglement and ensuring efficient evacuation. This D 83 command is indispensable for the efficient drilling of deep cavities when employing the GA 83 command. To regulate the Peg drilling operation, you must specify numerous parameters. Consequently, in order to employ the syntax, you must first enter d g 83 and subsequently specify the position of the perforations using the x, y, and z coordinates. The retract planes specified below are as follows. The break depth type is specified by you, while the fit rate is denoted by f. As illustrated in the figure, it is possible that we must construct a deep tunnel that extends to this point. Therefore, it is somewhat challenging to drill a cavity continuously until it reaches the bottom. Therefore, we can employ this Peg drilling cycle for this purpose. The drilling operation will be conducted in a step-by-step manner to create a cavity by a specific date. This implies that we will initially dig a trench to the specified depth. Again, the tool will be retracted to the reference plane. The drilling operation will be resumed once more.

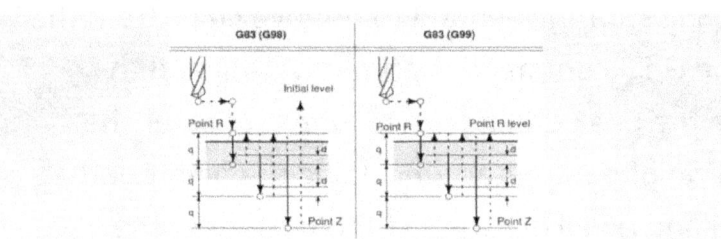

### G83 - Peck Drilling Cycle
- Used for peck drilling cycles, particularly for deep-hole drilling.
- It automates the drilling process with intermittent chip-breaking pecks and retracting the tool.

The drilling will now proceed to the second phase. The instrument will be retracted once more. Lastly, the drilling is conducted once more, and a cavity is generated to the necessary depth. Lastly, the instrument is retracted to its initial position. The instrument has now returned to the initial level, or the reference level. That is contingent upon the application of the G 98 or G 99. Therefore, the deep pit is constructed through a series of incremental drilling operations, which are feasible through the utilization of this Peg drilling site. Alternatively, as illustrated here, we have provided a single instance. So, initially, G 83. The coordinates are x ten, y twenty, and z fifty, with r being five, q ten, and f 50. Therefore, the CNC machine is the first to encounter the g 83 command. Rapidly position the instrument at the initial puncture location at x ten and y 20. The drilling process is subsequently initiated at d z -50. The number ten is used to represent the q, as is evident in this example. Therefore, the drill will be retracted to the designated peg depth during each peg cycle. The ten m m is responsible for breaking the fragments and averting their accumulation. The tool is retracted from the hole after reaching the total depth, and the down feed is controlled by the specified feed rate f of 50 during each peg, ensuring efficient material removal. In conclusion, the G 83 command is a decoder that is employed in the Peg drilling cycle of C and C machining. It ensures efficient Gpio activation, reduces chip entanglement, and maintains the quality of the drill hole by automating the creation of deep cavities with multiple pegs. We will engage in a more in-depth conversation regarding the critical CAD cycles during our subsequent decision.

# ADVANCE G CODES PART 5

Our discourse will commence with the subsequent critical G chord. The G 84 command is utilized in CNC machining to specify a tapping cycle, and it specifies the tapping and cycle D. So this tapping is the process of establishing internal threads within a pre-drilled cavity. The CNC machine is capable of producing threads by drilling holes as a result of these four commands. The primary objective of the GA for command is to automate the input operation. It enables the specification of parameters, including the tempo, input rate, spindle speed, and depth, which guarantees the precise and consistent creation of threads. It is utilized in the same manner as the drilling cycle; however, the tapping instrument will be employed in place of the drill. When employing the d8 command. The Tap feature is created using the G 84 command, and this command is essential. In order to utilize this G 84, it is necessary to initially tap it, as you specify a number of parameters to regulate the tapping operation. The host position is then specified by supplying the x, y, and z coordinates. Once more, R denotes the retract plan, while

F denotes the fit right. This is an example of the reference we have provided. Therefore, the tapping operation will be executed by the CNC machine when it encounters the PG 84 command, as is evident. Initially, it will swiftly relocate the tool to the initial location, which is x ten and y 2010.

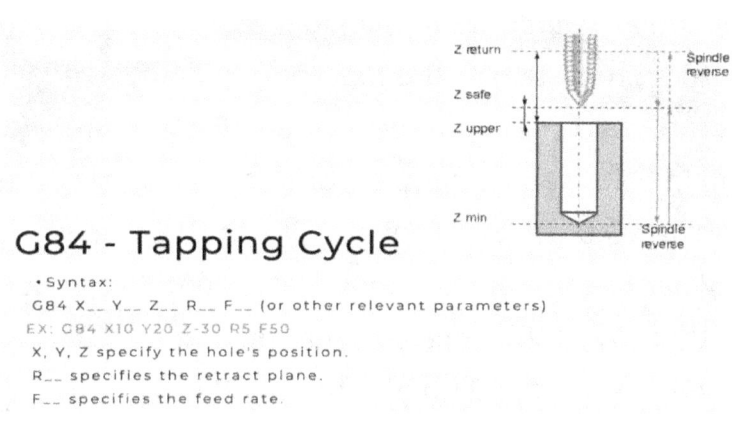

### G84 - Tapping Cycle

- Syntax:
G84 X__ Y__ Z__ R__ F__ (or other relevant parameters)
EX: G84 X10 Y20 Z-30 R5 F50
X, Y, Z specify the hole's position.
R__ specifies the retract plane.
F__ specifies the feed rate.

The drilling operation will be initiated at z -30. The downward movement of the tap as it penetrates the workpiece is regulated by the feed rate f 50. Once the tap has reached the designated depth, it commences threading. Operation. The rotation of the touch tool to generate the tag is regulated by the spindle speed, which can be configured in a different location within the program. This G 84 is especially advantageous for applications that necessitate the creation of precise threads within pre-drilled holes, as it is retracted from the hole at the designated distance after the threading process is finished. For example, fasteners or bolts. These specific parameters in the G 84 command may differ depending on the CNC machine and the tapping requirements. The precise and reliable threads can be achieved by ensuring that the proper configuration and tool selection are made, as tapping operations can be sensitive to factors such as the material, spindle speed, and feed rate. In conclusion, the G-code 84 command is employed in CNC machining to automate sealing cycles. It increases the accuracy and repeatability of the process by specifying parameters such as input rate, depth, and tool position, thereby simplifying the creation of internal threads within pre-drilled holes. The GA 88 is the next significant standardized cycle, which is the boring CAD cycle. The DD 88 command is used in CNC machining to perform precise and boring operations. It automates the voting process, which encompasses the rapid positioning, piercing, and retracting of the tool at the designated length. The boring operations are simplified and standardized by the GA eight that you have established. So, these boring is the process of enlarging or refining an existing cavity in a workpiece to attain precise dimensions. The GE 88 command is primarily designed to automate the boring process, enabling the precision control of tool positioning and depth to ensure accurate and consistent full dimensions. It is particularly well-suited for applications that necessitate close tolerances. Several parameters are specified to regulate the operation when employing the G 88 command. This is also the case with the syntax. Initially, the PG 88 command must be employed to specify the position of the openings. The d, x, y, and z coordinates are specified in the same manner as in the previous flame cycle. The filter is specified by the retract plane and f. Here, the modifications were limited to the replacement of the drill tool or the tape tool. The boring instrument will be employed as a reference. The CNC machine will execute the boring operation by first swiftly positioning the tool at the initial hole location of x ten and y 20 and subsequently commencing the boring operation at the z -30 location when it encounters the "get eight boring cycle" command. The tool will then proceed to remove the material from the interior surface of the hole in order to accomplish the desired hole diameter. The downward movement and material removal are regulated by the feed rate 50. The instrument is retracted from the cavity upon completion of the operation. Upon the completion of the ultimate diameter. The instrument retracts from the opening that has been created after the operation has been completed. The DG 88 boring cycle is frequently employed when a cavity can be readily drilled with a high degree of precision. Dimensions, which guarantee that the boring instrument reaches the specified location with precision. The G 88 boring cycle's precise parameters may differ depending on the CNC machine and the boring operation's unique requirements. The precise boring results are contingent upon the proper selection of tools, alignment of the tools, and specifications of the materials. In conclusion, these are assigned eight tedious cycles. Is the G-code employed in CNC machining to automate precision milling operations? By specifying parameters such as tool positioning, final diameter, and feed rate, the process of enlarging or refining openings is simplified with high accuracy and repeatability. In the same vein, the CNC machine offers a variety of preset cycles; however, their implementation is nearly identical. As you are now cognizant of these, some of the critical inputs, such as the selection of the retract plane and the provision of the pitch or depth, are essential for utilizing the various preset cycles available in the CNC machine. Now, the G80 is the next critical code. The G80 command in CNC machining is used to cancel a canned cycle or a fixed cycle, which is a predefined sequence of decoders that automates common machining operations such as drilling, tapping, or loading. Additionally, it indicates the cancellation of any of the selected cycles.

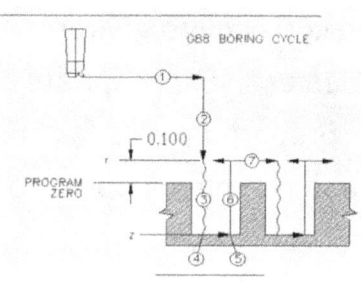

## G88 - Boring Cycle
• Syntax:
G88 X__ Y__ Z__ R__ F__ (or other relevant parameters)
EX: G88 X10 Y20 Z-30 R5 F50
X, Y, Z specify the hole's position.
R__ specifies the retract plane.
F    specifies the feed rate

These G80 commands are used to terminate or cancel such cycles. Therefore, this command is typically employed to terminate a can cycle. The process of restoring the machine to its normal operation has been initiated. Additionally, it guarantees that no cycle is currently underway. A DJT command is employed to terminate a specific cycle and may be positioned at any point within this CNC program to facilitate a subsequent exit. Therefore, in order to employ these, one must merely enter the letter "a." Therefore, whenever the CNC machine encounters this command, it terminates any active cycle and reverts to its standard operating mode. This enables the execution of additional non-canned cycle commands or the transition to a different section of the machining program. The G80 command is employed to transition between various machining operations, particularly when transitioning from scheduled cycles to other types of commands. The G80 command must be placed correctly to guarantee that the scheduled cycle is canceled at the appropriate position in the program. These prepared cycles facilitate the programming of common machining operations, thereby reducing the risk of errors and saving time. The use of these Gat commands is again essential for the maintenance of the program flow. So far, we have addressed the significant advance, the G codes, and a few of the most significant types of icons, as well as the process for canceling them.

## ADVANCE M CODE

These are a few of the most significant advanced M codes that we will be discussing. Therefore, we will commence with the initial one, which is the M ten. Additionally, I am eleven years old. It also indicates that the pallet clamping and on clamping m ten and m 11 are employed in CNC machining centers with pellet changers. So, m ten is employed to claim the pellet in its current location, while m 11 is employed to unclamp the particle for removal or exchange. Therefore, the specific functions and devices that are regulated by m ten and 11 may differ among various CNC machines. They are generally linked to the fastening and securing of workpieces or fixtures during the machining process. The utilization of m ten and m 11 commands is discretionary and contingent upon the requirements of the machining process. These commands will not be necessary for all CNC devices or processes. The correct and effective use of m ten and m 11 commands is contingent upon the proper documentation and configuration of this CNC machine. In conclusion, the m ten and m 11 commands in CNC machining are employed to regulate auxiliary devices, which are frequently associated with the fastening or securing of workpieces or fixtures. The specific function of M10 and M11 is contingent upon the machine's configuration and the machining operation's requirements. M10 opens the clamp, while M11 closes it. Consequently, in order to employ these functions in a straightforward manner, it is necessary to enter either m ten or m 11. The clamp is opened by the CNC machine whenever it encounters the command M10, and it is closed by the CNC machine whenever it encounters the command M11. This command is utilized to regulate the spindle and coolant supply, as the M 13 is the subsequent critical m core.

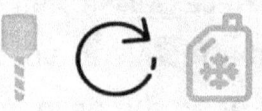

## M13 - Spindle On, Clockwise with Coolant On

- M13 is a combination M-code that starts the spindle rotation in the clockwise direction and activates the coolant system.

Therefore, the spindle and coolant system are activated by the m 13 command. Therefore, these m 13 are a combination of M code that initiates the spindle rotation in a clockwise direction and activates the refrigerant system. It is employed when refrigerant is required during the spindle rotation. Therefore, the primary objective of the m 13 command is to prepare the machine for a machining operation that necessitates the coolant system to be active and the spindle to rotate in the clockwise or forward direction. This is typically employed during cutting operations when lubrication or chilling is required to prevent tool VR and overheating. Therefore, the CNC machine will activate the spindle, determine the direction of rotation, and, if configured appropriately, activate the cooling system when it encounters the M 13 code. This ensures that the tool is correctly lubricated or chilled during the process and prepares the machine for the cutting operation. Therefore, in order to employ these, it is sufficient to input M13 into the program whenever necessary. The configuration of the machine and the specific requirements of the machining procedure determine the use of M13. It is not necessary for the spindle to be operational and the coolant to be operational in all machining operations. In conclusion, the M13 command is utilized in CNC machining to activate the spindle in a clockwise direction and activate the lubrication system. It is necessary to prevent overheating and tool fatigue during the machining process. Constituent. The M14 is the next significant code. It is similar to the M13 in that it activates the refrigerant system and turns on the spindle in a counterclockwise direction.

M14 - Spindle On, Counter clockwise with Coolant On

- M14 is a combination M-code that starts the spindle rotation in the counterclockwise direction and activates the coolant system.
- It combines spindle rotation and coolant application.

The M14 command's primary objective is to configure the machine for a machining operation that necessitates the coolant system to be operational and the spindle to rotate in an anti-clockwise direction. In certain machining operations, it may be necessary to rotate the spindle in a counterclockwise direction. The M14 command is inserted into the CNC program to initiate the spindle in a reverse direction and activate the coolant. If the CNC machine is configured correctly, it will activate the spindle, rotate it in a counterclockwise direction, and activate the coolant system when it encounters the M14 command. This prepares the machine for the cutting operation by ensuring that the tool is adequately lubricated or frigid during the process, while the spindle rotates in the opposite direction. The configuration of the machine and the specific requirements of the machining procedure determine the use of M14. For specific machining operations, it may be necessary to reverse the spindle. In summary, the M14 command in CNC machining is utilized to activate the spindle in a counter-clockwise direction and activate the coolant system. This prepares the machine for cutting operations that necessitate spindle reversal and lubrication or cooling to prevent overheating and tool wire overheating. Therefore, the M13 and M14 are utilized in a similar manner; however, the M13 is advantageous when necessary. When it is crucial to have a counterclockwise rotation, M14 and clockwise movement are necessary. The M 98 code is the next critical M code, and it is employed to invoke the M 98 command in the subprogram. CNC machining is employed for the invocation of macros or the summoning of subprograms. It directs this CNC machine to execute a predefined subprogram or macro program during the execution of the primary CNC program. This program is a valuable instrument for the simplification and repurposing of code in CNC programming, as it can include a sequence of decodes and other commands.

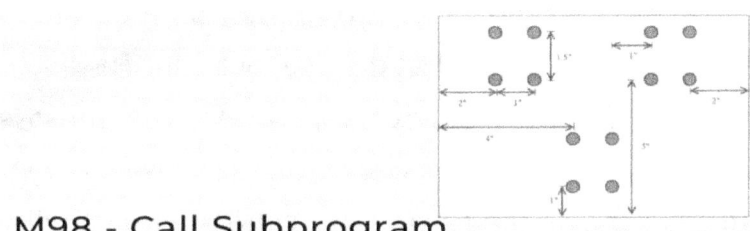

## M98 - Call Subprogram

- M98 is the M-code used to call a subprogram from within the main program.
- It initiates the execution of a separate, predefined subprogram at the specified location.

The primary objective of these M 98 commands is to modularize a CNC program. It facilitates the organization of the code into a logical sequence, simplifies program maintenance, and enables the reuse of the code. The programs are frequently employed for repetitive tasks, tool changes, tool offset adjustments, or other operations that occur frequently. The M 98 command is employed in the primary science program to invoke a specific subprogram. The primary program is temporarily suspended, and the specified program is initiated. Therefore, as evidenced by the current syntax, the initial step is to input M 98, which is followed by the letter P, which denotes the program number. Therefore, if we have entered the value of m as £0.98 1001, the program named 1001 will be executed. And after it is finished, the machine returns to the primary program at the point where the M 98 command was executed. Once more, the CNC program proceeds to execute from that precise location. The utilization of M 98 is highly advantageous for the purpose of streamlining CNC programs, facilitating the reuse of code, and enhancing the organization and maintainability of the programs. The programs can be defined either within the primary CNC program or externally in distinct files. The M 98 command is employed in CNC machining to invoke subprograms and to execute predefined subprograms or macroprograms during the execution of the primary CNC program. In summation, this command. It encourages recycling. It streamlines the programming process and arranges the code into modular segments. The subsequent significant code is M 99, which is employed to denote the conclusion of the subprogram. Therefore, these M 99 are employed to indicate the conclusion of this program. That is referred to as employing the M 98. In conjunction with these programs or the macro programs, the M 99 command is employed in CNC machining. It is employed to transfer control from the subprogram to the main program subprograms, or more specifically, the segments of C code that can be programmed from the main program using the M 98 command. The M 99 command returns the C machine to the main programs after the program has been executed. Execution. The primary objective of the M 99 command is to transition from a sub-program to the main program, thereby guaranteeing that the CNC machine resumes the execution of the main program from the point at which it left off. The M 99 command is employed within the sub program to signal the end of the sub program and to return control to the main program. In order to execute these M 99 commands. It is sufficient to input m 99 into the subprogram, as evidenced by the initial return of 01001 in the example.

### M99 - End of Subprogram

- M99 is the M-code used to indicate the end of a subprogram.
- It is used to mark the completion of a subprogram called with M98.

Therefore, the designation "100" signifies the commencement of this program. The g zero is the positioning command that is included in this program. The G one command commences a cutting operation in the main program, while the M 99 command is used to terminate the sub program. Therefore, the CNC machine terminates the subprograms whenever it encounters the M 99 command within a subprogram. Execution and subsequent return to the primary program. Therefore, the utilization of M 99 is indispensable when employing subprograms to guarantee a seamless transition from the execution of the subprogram to the main program. The M 99 command is employed in each sub program to guarantee that control is returned to the appropriate location in the main program, as sub programs can be invoked multiple times within the main program. The effective programming of CNC machines and the utilization of the M 99 and M 98 commands are contingent upon the proper labeling and organization of subprograms. In conclusion, the CNC programs' efficient organization is contingent upon the M 98 and M 99 commands. M 98 enables the summoning of sub-programs to simplify and modularize C and C code, while M 99 is invoked within these programs to guarantee a seamless transition back to the main program. Execution. The code is promoted, reused, S programming is simplified, and the CNC program maintainability is improved by this combination of commands. Thus far, we have addressed several critical M codes. We will utilize these various M codes in our subsequent session and gain a deeper understanding of the subject matter by conducting practical simulations.

# COMPONENT 1

We will endeavor to develop the novel component. Additionally, as illustrated on the screen, our objective is to generate two files that contain gaps in specified locations. Additionally, each of these openings has a radius of five. These openings are available at a distance of ten from each end point, and they all have the same dimensions. Therefore, we can replicate the entire creation process for each and every cavity. Therefore, we will endeavor to utilize this program and subsequently recollect it for the purpose of drilling. Therefore, we should proceed to the CNC Simulator Pro. Our goal is to generate five distinct openings, as indicated on the screen. Additionally, each of these openings has identical dimensions and depth. Therefore, we will endeavor to leverage this program. So, the code to comprehend that has already been provided on the right-hand side. Initially, our goal is to load the machine. The milling machine is the machine that is necessary in this instance.

Therefore, the machine can be loaded by accessing this file and selecting the "load machine" option. Or, you may chose to access the machine by selecting the "open" option from the bottom panel. The workpiece will be selected subsequent to the machine selection. This workpiece has a dimension of 80 by 65 by ten. This implies that the workpiece is 80 inches in length, 65 inches in breadth, and has a retained depth. Additionally, the workpiece has been incorporated by employing the Ada resistor line roller, as evidenced by the initial line.

The index of the workpiece is denoted by part eight. This index may differ in your specific situation. Subsequently, I have determined that the workpiece should be positioned at a distance of 30 in the x, y, and z directions from the machine zero. Therefore, we should expand our perspective. This is our machine zero, as you can observe. Therefore, I have positioned this workpiece at a distance of 30 NM in the y direction, 30 m in the x direction, and 30 m in the z direction. All dimensions in millimeters will be taken into account in this section. Therefore, we have composed the following: d g 21. To utilize the metric input. Next, the entire machining process will be conducted exclusively on the x-y plane. So, we have chosen 17 to consider the absolute programming, and G 90 is selected. G 94 is used to consider the input in M per minute after the workpiece has been set. The set-up of the instrument is our obligation. Return to the geometry once more.

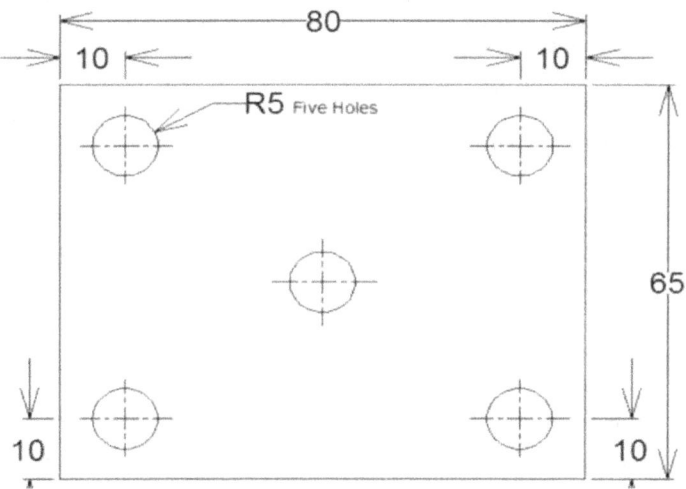

Our objective is to establish the cavity, as evidenced by the image provided. And these openings have the appropriate radius. Therefore, it is imperative to choose the instrument with the appropriate radius. Therefore, we have chosen the 8017 for this purpose. Additionally, this 80 represents the embedded instrument. The radius is accurate in the 8017 index utility that is currently available. Therefore, I must now provide the dimension from this apex for any necessary machining operations after selecting the tool. Therefore, I will establish the program's zero point. At present, this program zero is accessible from a distance of 30 from the x-axis, 30 from the y-axis, and 30 plus its depth. One million meters. Therefore, the value of m doubled from this machine is zero. Therefore, I have chosen d g 92, x 30, y 30, and z 30. Thus far, we have adapted the workpiece and d tool to meet our specifications. As you can see, our objective is to generate these openings, each of which is situated at a distinct x- and y-coordinate. However, it is evident that the spindle must be activated whenever the instrument approaches this particular position. Afterward, it is necessary to activate the refrigerant supply. Finally, in order to account for the dip point of the D drill, we will be supplying an additional dimension of d depth ten M to create a perfect cavity. Therefore, we will endeavor to provide the profundity from the top to the 15 below. Therefore, we will furnish information relative to the subsequent generation of -15 M. The spindle will be timed. Finally, the instrument is retracted from the cavity.

And once more, deactivate the refrigerant. Therefore, this advice will continue to be applicable to all of these openings. Therefore, we will endeavor to develop this subprogram for this particular operation. Consequently, we have developed this program, as evidenced by the right-hand side. We will commence the numbering of this program at the beginning. Subsequently, the program number is determined by the number we have assigned. Therefore, we initiated the program with the number 01234, which signifies that it is currently in progress. We will endeavor to execute all necessary operations within this. Therefore, the spindle is activated in a clockwise direction at a speed of 1200 rpm. Next, coolant is activated if the value is 0.05. Ultimately, we will attempt to generate the crater by employing linear interpolation, with a depth of -15 MB. Initially, we will attempt to retract the tool from the workpiece after it has been created. Therefore, we should initially maintain this line from G00 to z three. Following the tool's withdrawal from the workpiece. Lastly, the spindle will be prevented from rotating. Therefore, I will press the "m" key to halt the spindle. The refrigerant is turned off by the subsequent M09. Lastly, it is imperative that we provide the eight plus m 99 to ensure the program concludes. So, the final command, M 99, is crucial for the transfer of program automation back to the primary program. Therefore, we will invoke this program 1234 whenever necessary to execute this particular operation. Therefore, as you can observe, we will attempt to swiftly multitool at the initial position, which is the x and y coordinates, after setting the program zero point. Additionally, it is being maintained at a distance of six megabytes from this. What is this? Upon reaching this position, it is necessary to construct the cavity. Therefore, we will be contacting 31234. Therefore, in order to invoke this program, we will employ the command M 98 and the program number, followed by d p. That is the number 1234. Therefore, it will execute the machining operation by returning to this program. Additionally, this initiative persists. Ultimately, it conducts the M 99. Consequently, the primary program continues to operate while certain programs are terminated. Once more, we will expedite the process of moving the instrument to the subsequent position. Therefore, it is evident that we are offering x at 70 and y at ten. Additionally, we will execute the drilling operation once more with the assistance of this software. In the same way, it is evident that we were compelled to swiftly relocate the tool and subsequently execute the drilling operation using these subprograms for all necessary openings. The number 21234. In order to comprehend this operation, we will initially attempt to execute this simulation in order to provide you with a more comprehensive understanding. Therefore, I will select this play simulation. As you can observe, the sub program is invoked each time M 98 executes. These can be observed on the. I will also relaunch the program and examine the right-hand side editor. Additionally, the manner in which this line is executed in the CNC editor. Therefore, I will once more select this simulation. In order to comprehend the rapid movement and the freedom movement, we should examine weights and numbers. I will now select the plane simulation. The workpiece was subjected to repetitive traverses, as is evident. Obtain the appropriate machining. Therefore, it is evident that the duration of our CNC program decreases when this program is implemented. And we can attain the accuracy and repeatability. Therefore, this is the ultimate component that is generated by this software.

# COMPONENT 2

We will endeavor to generate this geometry, and in order to do so, we will implement a tool radius compensation.

Therefore, capture a screenshot of this geometry. Therefore, you possess a more comprehensive understanding of the various dimensions while programming. As demonstrated here, the standard dimension will always be measured from the tool's center point to the subsequent dimension. We will consistently relocate these center points to this other body. Therefore, let us assume that I provide this datum coordinate from this starting point. The focal point of these two then shifts to this particular data point. Therefore, it is evident that the circular tool reaches this datum point whenever a specific circular component is present. Subsequently, it automatically eliminates the surplus material that is of the same radius. Consequently, this results in machining that is inaccurate. Therefore, in order to comprehend this, we will navigate to the CNC Simulator Pro located on the right-hand side. As you can observe, the initial line is the workpiece that is accessible at the ninth index in this milling center. The code has already been supplied. This workpiece is situated 20 NM from the machining zero point in the x, y, and z axes, and ten m from the z axis. The workpiece we are utilizing has a land area of 100 m m in the x axis, a width of atm in the y axis, and a depth of ten m m in the z axis. This G90 configures the machine to operate in absolute positioning mode, which interprets coordinates as absolute positions from the machine's reference point. The equipment is configured to utilize millimeters as the unit of measurement by G 21. G 94 configures the machine to operate in units per minute or iterates. The G 92 command is beneficial for the vertical coordinate offset, as it establishes the coordinate system's origin at x = 10, y = 10, and z = 20. In other words, all subsequent coordinates are relative to this new object. The subsequent line, M06, is utilized for tool change and specifies that tool 81, v ten, and mil of six M diameter are to be employed. The spindle is initiated in a clockwise direction by the subsequent line M03, which sets the spindle speed to 1200 rpm and the feed rate to 20 units per minute. Next, M08 activates the connection. Therefore, we should initially establish the division piece and 22. For the time being, I will execute the code and select the simulation icon. Therefore, it is evident that this is the machine zero. The center point of the program is also provided. The coolant tool will be swiftly relocated to this 0Y0 after you switch it on. This implies that the instrument will be relocated to the subsequent line.

The tool will then travel downward from these reference points. So, it is program zero. The utility is currently accessible ten meters below this reference point. We will now proceed to the geometry, as evidenced by the start point or program zero. The tool must be relocated to the initial specific point, and the distance from the program's zero point to the workpiece's beginning point is ten meters in the x axis and NMM in the y axis. Then, linear movement is necessary. Therefore, it is necessary to supply t exclusively for the y-coordinate. After that, the x-y plane movement is once again necessary. Therefore, it is necessary to furnish the x- and y-coordinates. Once we have reached this particular stage, we must once again implement linear interpolation or linear trim. The clockwise circular interpolation is necessary, as one can observe in this image. Lastly, a circular interpolation is necessary to be performed anticlockwise from this point.

And from this lowest point, one can return to the starting point. Linear interpolation is necessary. I trust that you will be able to obtain all of the necessary coordinates from the beginning point to the end point using this geometry. We will now proceed to the CNC Simulator Pro, where you will observe that we will initially relocate the tool to the y-axis and extend it. Therefore, I will proceed with this simulation. Therefore, it is evident that the utility is accessible from this program, zero point. The simulation will now be resumed. And as you can observe, the center point of this instrument has been relocated to the end point of this workpiece. So, you can see that we have provided the coordinates x ten and y ten. However, this specific instrument is necessary due to the material of this workpiece having a radius. Given that this instrument has a radius of three. Therefore, the trim material is automatically eliminated as a result of its dimension when the coordinate is supplied. Therefore, in this instance, it is evident that the tool radius compensation is necessary solely to cut the necessary material. Therefore, I trust that you have gained an understanding of the significance of this tool radius compensation during the machining process. Let us initially proceed with our simulation using this specific code.

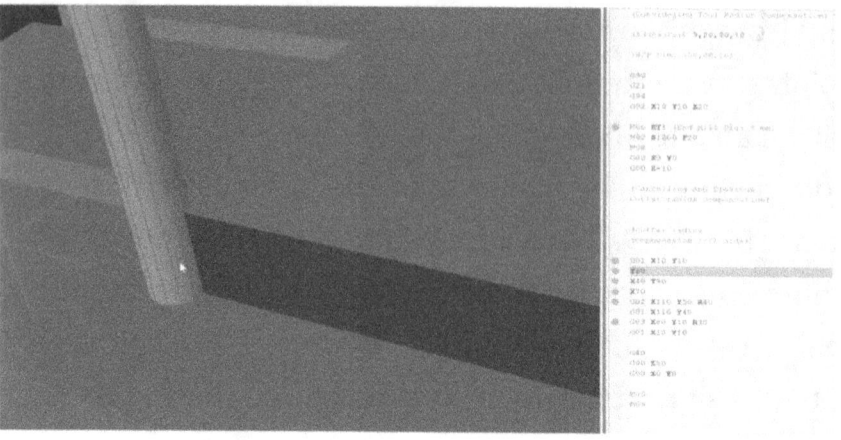

The instrument will now move by six centimeters in the y direction once more. The x axis movement is indicated by four tm, while the y axis movement is indicated by 90 m, as indicated by the point coordinates. Therefore, we will observe the simulation. Consequently, the three shift materials are mechanically removed by the tool when it moves in the x direction, as evidenced by its radius. So, I will once more proceed with the simulation. Circular and clockwise interpolation is now necessary from this point forward. Additionally, the coordinates are as follows: X110, y50, and a radius of 40. Therefore, we should proceed with the simulation. From this point forward, an anticlockwise arc must be established. Therefore, the G0 core is once again employed in this instance. Additionally, point coordinates are furnished. That is the case. The y coordinate is set to ten, and the x movement is required to be up to 80 m in absolute coding. In this instance, the radius is 30 meters. Therefore, we should proceed with the simulation. Finally, a direct copy has been produced and will be transferred at a rapid pace. Therefore, it is evident that the three-move material is removed in each direction by the tool if the cutter radius compensation is not utilized. An additional component is required in this instance. In this instance, it is not practicable to conduct a thorough machining operation. Therefore, we will implement tool radius compensation. Therefore, we will first cancel the entire prior tool radius compensation on the right-hand side, as you can see. Therefore, it is more practical to terminate the tool radius compensation at the outset, as we are uncertain whether any compensation was provided during the previous machining operation. Presently, afterwards. The tool is progressing toward the left side of the workpiece, as is evident. Therefore, in that scenario, it is necessary to provide tool radius compensation on the left and right sides, as well as on the left. The code that is necessary is dt 40. What is the matter? Therefore, I will employ the command D 41 in this instance. Therefore, it is evident that we have incorporated d g 41 to account for the left-hand side cutter radius compensation. In the end, we have once again canceled the tool radius compensation. Therefore, in order to obtain a more comprehensive understanding of the fields and traverse, we should initially examine the weights and bus drivers option. I will now resume the simulation. Please allow me to recalibrate the view in order to obtain a more expansive perspective. Right now.

I will resume the simulation once more. The linear interpolation will now be provided. We will now resume the simulation. Now, as you can observe, the tool is autonomously relocating the tool part VM from the original toolpath as a result of the cutter radius compensation. The reason for this is the utilization of tool radius compensation. Once more, we will proceed with this simulation.

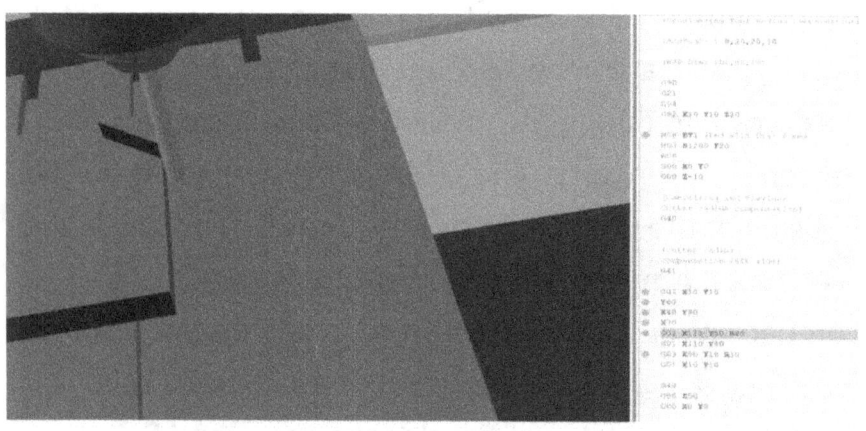

I will initially reset the view, zoom in, and then proceed with the reoperation. Therefore, you can once more observe the three deviations from the original toolpath. Therefore, the excess material will not be eliminated. As you can see, the three M's are also present on this side, and they are displaced by three arms due to the tool's available radius. We should conclude this simulation.

Now, I will first eliminate all of the pose positions. Again, we will proceed with the simulation to obtain a more comprehensive understanding of the toolpath. Therefore, this is the geometry that we have established, as you can observe. Therefore, I trust that you have a clear understanding of the situation. The compensation of the tool radius is extremely beneficial for the execution of specific machining operations.

# COMPONENT 3

We will endeavor to generate the geometry as it appears on your screen. Initially, capture a screenshot of the geometry to facilitate your understanding of the various coordinates during the programming process. Therefore, we will proceed to the CNC simulator Pro located on the right-hand side. It is evident that the necessary code has been generated, and we will endeavor to comprehend each and every line that has been written. Therefore, the CNC simulator Pro command is the initial command and registry component. And it selects the tool that is available at the D11 index, and it sets the coordinates to D 15 in X, 15 in Y, and ten for z. The machine is configured to operate in absolute positioning mode by G90, which interprets coordinates as absolute positions from the machine's perspective. However, G 21 instructs the machine to utilize millimeters as the unit of measurement. The x-y plane is chosen for machining by G17. It suggests that the CNC machine will function in the x-y plane. The machine is configured to utilize units per minute for the input rate using G90 four. The machine was configured to utilize units per minute for the fitted command by G 94. A word coordinate offset is defined by A G 92. It maintains the coordinate system's origin at x=0, y=0, and z=20. It specifies the use of tool 81 with a diameter of six M, which is the embedded tool that is already available in the simulator. The subsequent M06 command modifies the tool. The spindle is initiated in a clockwise direction by the Pro M03 command. This entails a maximum speed of 1200 rpm and a capacity of 200 units per minute. M08. The coolant is activated, and the subsequent line denotes the swift positioning to the necessary coordinates.

Therefore, we should execute the simulation. Up until now. Therefore, the workpiece has been incorporated, machine zero is accessible, and the program zero point has been included. It is once more time to return to geometry.

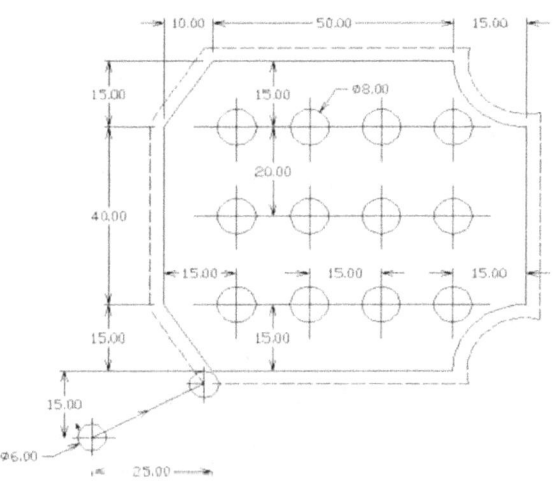

Therefore, as demonstrated by this program, zero one fragment is accessible 15 MB away in the x axis and 50 NM away in the y axis. As a result, the program zero specified that eight positions were necessary. So, as you can observe in the CNC Simulator Pro, the program zero is 15 MB distant from this specific workpiece in the x and y directions. Once more, we will execute the simulation. Subsequently, we must implement linear interpolation and rapid positioning to generate the necessary geometry. I trust that you now have a clear understanding of how to provide the coordinates for the machining operation. And in this instance, we have refrained from employing the cutter radius compensation. Therefore, it is imperative that we exercise caution when supplying coordinates for various locations. The initial step is to eliminate the tool's radius. I will resume the simulation after the machine is required to be run again. So, as you can see, the initial step. The G0 one code is employed to execute machining in the x axis. A circular arc is now necessary, and it must be in the clockwise orientation. Therefore, we have implemented the DG zero two command. Additionally, the radius values were taken into account, as well as the end coordinates of 90 and 27. Again, it is evident that a circular trajectory is necessary, and it should be in a clockwise direction. We have once again selected DG zero two and have successfully completed the remaining milling operation through this command. Therefore, we should proceed with the simulation. Let us revisit this geometry after the end milling is finished. Additionally, it is imperative that we generate these numerous gaps, as they are necessary. The guide cycle is a practical tool for performing repetitive operations, as it is highly beneficial. Additionally, this is merely a drilling operation. Therefore, we will implement the piercing and cycling processes. Therefore, let us once more visit the CNC simulator. Therefore, it is evident that each opening has a diameter of eight millimeters. First, we will replace the tool and select a drill tool with a diameter of eight MB. Therefore, we have implemented the M06 command. The embedded tool at the 16th index was selected in this instance. Additionally, it contains eight megabytes. And then, once more, we have swiftly relocated the instrument to the necessary location. Next, activate the spindle. Subsequently, linear interpolation is necessary in each drilling operation. So, initially, we have chosen the Z01 command. Subsequently, we wish to reiterate the initial level a tad following each drilling operation. Therefore, we have chosen the G 98 command to activate D drilling and cycle. The D G 81 command was implemented. The first hole's coordinates are d x 30 y 30. Then, drilling is conducted at a distance of d minus 2LMM to create the cavity.

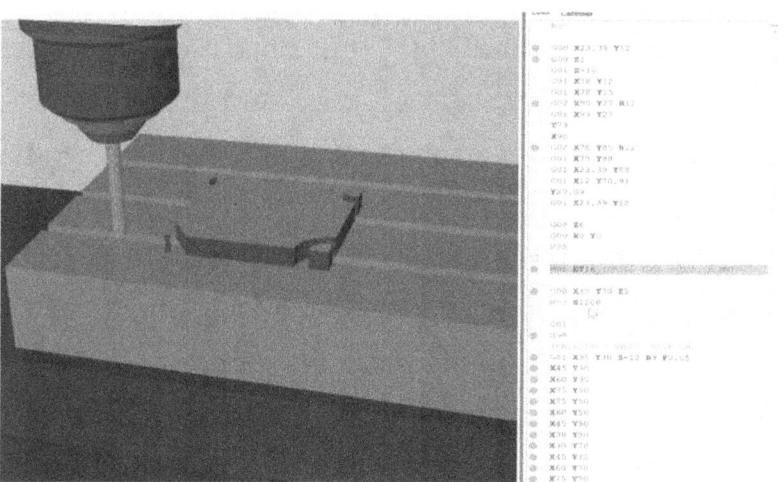

The plane is provided at r t, and the feed rate is set to 0.05 units per minute. This is to be reiterated. Therefore, we should initially proceed with these simulations. Therefore, as you can observe, the instrument has been modified. The utility is integrated. Next, we have chosen G 98. This implies that the instrument will be restored to its initial state. The simulation will now be resumed. So, as you can observe, the initial cavity is formed, and the tool is retracted to observe the appropriate drilling process. We should activate the D inputs and begin the traverse, as we have already dug the initial trench. At this time, we will furnish the coordinates for the adjacent location. Therefore, this is equivalent to d x 45 y 30. Once more, proceed with the simulation. Therefore, it is evident that we have merely furnished the coordinates. However, the drilling operation was also autonomously executed and the tool was retracted to its initial position as a result of the flame cycle. Let us once more inspect the subsequent cavity. Again, it is evident that D drilling is implemented, and the instrument is detected at its initial position. Consequently, the grid is established by providing the subsequent drilling coordinates in the same manner. Therefore, we should proceed with the simulation for all remaining openings. Finally, the G8 command is employed to conclude the drilling cycle, and the tool is redirected away from the workpiece after all the holes have been created at the designated locations. To ensure that the entire operation is correctly visualized, it is necessary to first eliminate all the elements after the disturbed quality of the and d program is resolved. I will resume this simulation after removing all the post locations. Therefore, examine the left and right sides. Therefore, the manner in which the movement occurs is contingent upon the various positions. Therefore, I will increase the magnification to obtain a more comprehensive perspective. I will now proceed with the simulation. So, the initial step is to perform the milling operation. Additionally, the necessary outer geometry is established. The instrument will be modified at this time. Drilling tool was selected, and the CAD cycle was continued. I trust that you have gained an understanding of the significant benefits of these cycles in the context of repetitive duties. Therefore, we have implemented the drilling and cycle in this instance. The taping cycle and piercing cycle can be implemented in accordance with your specifications in any of these CNC machining programs.

# COMPONENT 4

We will comprehend the utilization of the CNC simulator programming machine. To achieve this, we will generate the geometry that is visible on your screen. It is important to note that all dimensions are accessible in each, and we will be using the radius as our input command for machining purposes. Let us first navigate to the CNC Simulator Pro. Therefore, as evidenced by this image, the milling machining center is accessible. First and foremost, it is imperative that we identify a profound learning center. Therefore, navigate to the bottom panel and select "Open Machine." Select "OK." Additionally, this is where you will locate the option to enable double-clicking. Additionally, choose the "Turning Center" option. Uncheck the "Open Demo" option. Next, select "Open." The CNC Simulator Pro is currently equipped with a turning machine, as is evident. So, initially, it is not the case that we are employing the IT unit. Therefore, in order to configure it, you must initially navigate to the settings. Click on the settings button once more. Additionally, ensure that the units you have chosen are utilized in the subsequent term of the entire program. Please ensure that this option is checked, as we will be employing the radius coordinates for the machine. That is, the legs are to be positioned using the radius coordinates. If this option is unchecked, the CNC Simulator Pro will interpret all input coordinates as diameter coordinates. Now, select "OK." We are now required to develop a workpiece for our endeavor. Additionally, I have already generated the necessary workpiece. Therefore, in order to generate the workpiece, navigate to the parameters. Once more, select the Inventory Browser. Proceed to the LED. In your situation, which measure should be taken to incorporate the workpiece?

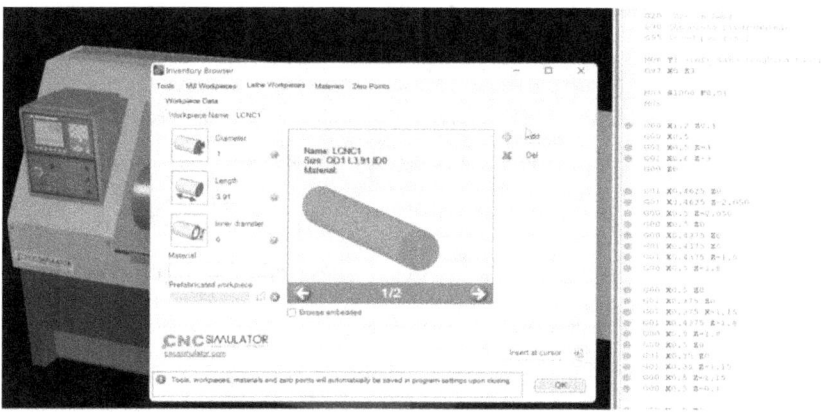

To add a new workpiece, first click on the "add" icon. The workpiece name must be entered after clicking on it. In my circumstance, I have entered d l cnc one. Afterward, the diameter must be given. In this instance, the necessary units must be supplied exclusively in accordance with the units that have been chosen. TLC Simulator Pro will automatically recognize all of these units as H or D millimeters, regardless of whether they are H or m. So, our diameter will be one h in this case. Therefore, I submitted one. Subsequently, enter D 3.91 as the length and zero as the inner diameter. Given that you have submitted all necessary parameters, it is imperative that you examine the index of this workpiece. These indexes may differ depending on your usage. Additionally. To incorporate this workpiece, simply click on the insert located at the cursor. I have already incorporated these in my situation. Please allow my workpiece to be displayed in this editor, and I will simply select "OK." Consequently, we initiated the process by inserting the recently generated workpiece into the project. This identical command. You have the option to manually input or select the "insert" and "cursor" options. At this point, the index number of the workpiece is determined by the final number. Our subsequent suggestion is to choose an implement for our roughing incisions. I have provided specific inputs prior to selecting the instrument, such as the G 20 code for edge programming. The machine is configured to operate in absolute positioning mode by G90, which interprets coordinates as absolute positions from the machine's reference point. There is a dimension to utilize units, thanks to G95; however, we will utilize it on a subsequent occasion. To reintroduce the tools, navigate to the settings. Select the Inventory Browser. Click on the embedded LED tools in this tool section. As evidenced by the image, there are numerous small tools available. To verify the tool's various index positions, simply click on the blue arrow. Therefore, we will exclusively employ the initial index number in this instance. Therefore, selecting the necessary instrument from this index position. The subsequent step is to select the cursor and insert options.

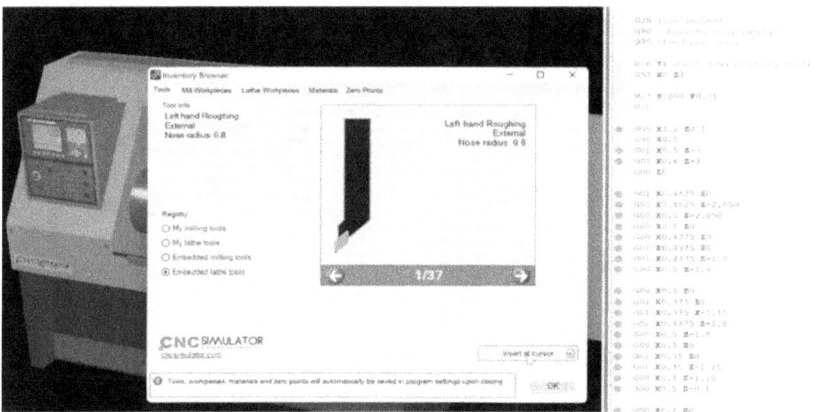

I have already incorporated the instrument in my situation. Therefore, I will select "okay." The M06 p1 has been selected, and the P1 denotes the left-hand roughing instrument. We are now prepared to commence the trimming of the workpiece. Initially, select the "reset view" option to enlarge the workpiece and ensure that it is visible in the bottom panel. We will once more take a closer look. We will now initiate the simulation to verify the inclusion of the workpiece and instruments. Therefore, I shall initiate the simulation. The workpiece is inserted in this task, and the D tool is also available. Again, we will return to the subject of geometry. Therefore, it is evident that the program zero point is accessible at this beginning position. Additionally, this tool reference point is accessible outside of this program zero. It is crucial to acknowledge that CNC Simulator Pro automatically positions the zero point in the correct plane of the jaws prior to commencing machining.

**Important!** CNC Simulator Pro automatically puts the zero point at the right plane of the jaws. The distance from the plane to the spindle is 23 millimeters. That means that the right end of our 100 mm long workpiece is at (100-23) 77, in the Z-axis. This rule goes for all lathes in CNC Simulator Pro. Always take the 23 millimeters into account!

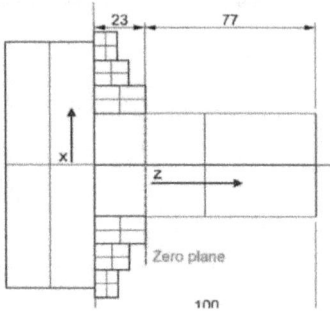

The spindle is 23mm from the plant. This implies that the right end of our 100-foot workpiece is located at 77. On the Z-axis. Therefore, as illustrated in the accompanying figure, if we utilize a 100 MB workpiece, 23 NM of that particular workpiece will be occupied within the machine, allowing us to work with only 77 MB of the workpiece. Therefore, this regulation applies to each stratum of this CNC simulator. In favor. Therefore, it is imperative to consider the 23mm when conducting machining operations. In our case, we have taken into consideration the need to provide an explanation, and to do so, we have utilized the length of D. T is 3.91 inches. However, let us revisit the CNC Simulator Pro. To begin the machining process, it is important to note that the coordinates provided with the X symbol always represent radial movement, while the coordinates provided with the d and z symbols always represent the length of the workpiece. Therefore, the z axis represents the length of the workpiece, while the x axis represents the diameter or radial movement. Within this object. Therefore, we should immediately transition to the instrument located at d x point five. Therefore, we should proceed with this simulation. Currently, it is evident that these workpieces have a diameter of one. The radius of this workpiece is 0.5. Additionally, let us further refine our focus. The program zero has been established at the beginning point of this workpiece, as is evident from this point. We should once more expand out from this location. Subsequently, subtract three from 301, which is equivalent to 0.50. Specifies that the instrument will be capable of moving within a radius of 0.5 and a maximum length of three eights. And since this distance is available in the negative direction from the zero point of this program, it is denoted by d minus three. Therefore, I will persist with this simulation. We will execute the subsequent line once more. In this instance, return to the geometry. The initial step is to adjust the dial until it reaches 95 degrees. And since we are providing the radial coordinates, it must be divided by two. Therefore, please be advised that the entire turning process will be conducted from the outer to the interior side. In order to preserve the geometry of this component, we will be supplying the radius as the coordinate. Return to the CNC Simulator Pro once more. We should conduct the simulation once more. Initially, we will increase the radius to 0.4625 and the length to D minus two. Each. Please return to D. Right now, we will assume that the radius is 0.4375. Therefore, the simulation will be repeated, and the turning process will be completed until the length of 1.80. Therefore, the simulation will now be resumed. Once more, the instrument will be redirected away from the workpiece. Return to the geometry. We have provided the machining from the dial point 75 to the dial 0.875, as visible up to the length of 1.8. The tapering operation is necessary. Additionally, it is necessary to furnish the precise coordinates for this purpose. Once more, navigate to the CNC Simulator Pro, where each initial turn will be executed for a radius of 0.375 and a maximum length of 1.15. Therefore, we should proceed with the simulation. The tapering will now be available from a diameter of 0.4375 to a length of 1.8. Therefore, the simulation should be resumed. This is where you can see that we have included the exhaust. The instrument is then retracted once more. The determination is now conducted at a radius of 0.35 and a length of 1.15. Once more, the utility is redirected. Return to the geometry once more. At the initial point, it is evident that the radius of point one requires a diameter of point five for each point. Therefore, we will initially necessitate an anticlockwise movement. Therefore, we have chosen DG03. The end coordinates are d point one, with a radius of 25 and a length of 0.1, the radius that is required. We will now execute this particular operation. Therefore, the simulation should be resumed. The turning process will now be conducted to a length of 0.65. Each. Subsequently, we require the taper. Therefore, coordinates are once more computed. Additionally, these coordinates can be determined by employing the trigonometric principles. Therefore, this simulation should be continued once more. Finally, we will halt the spindle rotation, deactivate the coolant, and unload it. Therefore, it is evident that we have developed the necessary component. I trust that this information has provided you with an understanding of how to perform determining operations that involve radial and longitudinal movement. I will commence the simulation after removing all of the previous push points in order to eliminate the entire component of machining. Therefore, examine both the left and right sides to ensure that you comprehend the concept. How. The entire shifting operation is executed by conducting this step-by-step operation. You acquire a more comprehensive understanding of the disciplines and traverse. Let us verify the boxes for rivers and vehicles. I will now select on the play simulation once more to examine this operation in order to provide you with a more comprehensive understanding. Therefore, I trust that you have gained an understanding of the process by which these machining operations are executed. Additionally, you have observed this phenomenon during the cutting process. We are employing DG zero one. However, we are employing the G00 for swift traversal during our return. Additionally, this reduces the amount of time required for machining. Consequently, these learning machines can be employed to execute any of the necessary rotating operations.

## CONCUSSION

We have reviewed several fundamental concepts, such as the concepts of T codes, codes to land offset, and told radius notions. With this information, you are on the path to mastering the craft of CNC machining. As you continue your voyage, it is important to remember that CNC programming proficiency is achieved through practice and practical experience. Avoid becoming disheartened by obstacles. Rather, consider them as opportunities for personal development and education. The manuals supplied by manufacturers of C and C machines are a valuable resource for further education. They frequently provide comprehensive information on G and M codes that are unique to their devices. Once more. Additionally, the CNC Programming Handbook by Peter Smeed is a highly esteemed resource for CNC programmers and operators. Another exceptional resource for acquiring a comprehensive understanding of CNC programming is Michael Madsen's CNC Programming Principles and Applications. For CNC software that is specific to your needs, it is recommended that you consult the official manuals and documentation. Another detailed explication of G and M codes that are pertinent to their systems is frequently included. The numerous community forums and discussion boards that are accessible on the online platform can be consulted. Because it is home to vibrant communities of CNC professionals and enthusiasts who exchange knowledge and experiences. The most effective method of acquiring knowledge about CNC programming and G and M code is through practical experience. Virtual C and C simulators are accessible, which enable users to simulate and practice programming without the necessity of a physical machine. These simulators are particularly beneficial for novices. Do not hesitate to experiment and attempt various code combinations. The most effective method of comprehending the complexities of CNC programming is frequently to learn by doing. Once more, you have the option to select a project that piques your interest and complete it from beginning to end. It is a fantastic learning experience to implement what you have learned in a practical endeavor, which could range from engraving to machining. CNC technology is constantly changing, so it is important to remain informed about the most recent advancements, software updates, and new codes or features that may improve your CNC machining capabilities. This will enable you to further your comprehension of the Recommandé art resources. Additionally, do not hesitate to seek assistance and direction from CNC communities and forums. Stay inquisitive and embrace the learning process, as DXF welding is a dynamic and ever-evolving process. The skills and expertise necessary to excel in CNC machining can be acquired through dedication and persistence. The capacity to program and operate CNC machines provides a plethora of inventive opportunities, regardless of whether one is a professional or a hobbyist. Continue to learn, experiment, and, most importantly, relish the CNC programming experience.

## INFORMATION OF POWERMILL

1. The topic of this lecture will be the inflammation of parliament. 2. Therefore, let us begin by simplifying certain aspects. 3 Like Parliament is a 3D game solution that operates on Microsoft. 4. We are aware that the instrument is programmed, but it is present for flight. 5 And appear to be willing to miss the upcoming

parliament's comfort and expert workshops in order to manufacture molds, dyes, and highly complex components. 7. Subsequently, the CNG devices can be optimized for efficiency and quality control. 8 Subsequently, it may be accomplished. The motion of five axes was validated and improved by the use of sophisticated simulation tools in a 9-quality box. I am not concerned with the software advantages of my scenarios, such as the optimal tool Leader Mine. However, I am interested in reducing the manufacturing process's time and the cost of the product. I am also interested in producing additional surface treatments for the company.

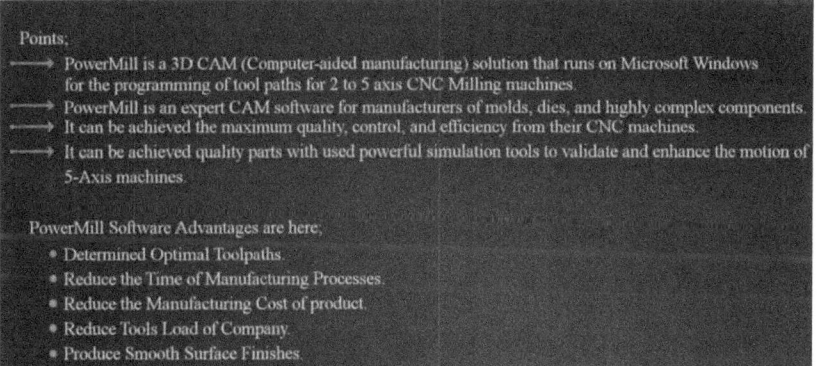

13 There are currently eight reasons to rapidly create a high-quality tool in software. However, on large and complex boxes, a high-efficiency rolling strategy is

necessary. The tool should be free and available in a variety of colors. The tool should be used for workweeks. The tool should be used for solids automotive programming, which includes an extensive library of do-or-die templates and macros. The tool should be flexible and can be used for editing, optimization, and automatic fire axis collision avoidance. The next benefit is the ability to handle even larger memory-intensive files, which is a benefit of the tool. It is four times faster as a tool, but the calculation takes a long time. The waiting time is reduced, and the tool's capacity for additional work is significantly increased. This improves manufacturing productivity.

## INTRODUCTION OF POWERMILL

1. The introduction of the power mill will be the subject of this lecture. 2. Initially, this is your PowerMill Softwear. 3. I have just opened it by double-clicking on it. 4. Consequently, PowerMill is currently operational. 5. You will observe that this is your fast access interface, which contains several options, including "open," "save," and "working plane." 6. If you add an additional option, click on it and deactivate it just like this. 7. The individual in whom you will perceive the logic in this situation is the same as the debtor. The tool path is not included in the suite of tools of 8. 9 Edit 2 settings boundary button hole feature feature grill work ordinary model stock model.

Simulate ten machine tools and observe the program view and Aditi. 11 Therefore, I would select a game. 12 This is simpler. Thirteen distinct. Fourteen distinct altitudes.

15 This is your explorer. 16 Therefore, in these, you will observe all the options for declaring our operation are available to you in the machine tool program. 17 Set a tool back tool boundary pattern feature group call feature set work simple label modern stock market. 18 Group and some simulation. 19 This is your current work. We are aware that this is your coordinate and that there is some underpinning here. 20 This is your new VQ. 21 This will alter your perspective on your object or workplace. 22. This is your toolbox; you will initially observe the meaning

of the machinery and 23. Dunning. 24 This is your view selection, such as the dog's front, rear, right, and left. 25 This is a distinct ISO view, such as ISO 1, 2, 3, and 4. 26 This is the final and simpler option.

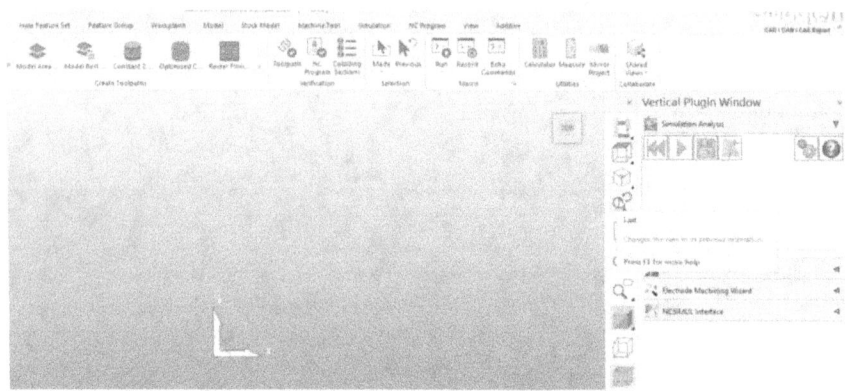

27 This is the pause view. 28 Zoom ft. Zoom by selection; otherwise, we determine your seating view through the free block selection matter and the previous selection. 30 This is your vertical plugin, and we are aware that the averages are close. I will now provide it to you. If you choose to select a walk plan, you will see it in your status column. Otherwise, you will be prompted to create a plan. 33 This is the plan you have chosen. 34 The x-y plane is now operational. 35 This is the Y for Z, which signifies that the plane is currently operating in this region. 36 This is the category you have chosen. 37 This is

the view of the X, Y, and Z coordinates. 38 This is the calculation that you require for your final perspective. 39 Ada clearance techniques are employed in this instance. 40 is the diameter of your instrument, and the final number is the radius of the indicator. 41 Friends, this is becoming increasingly bothersome. 42 The screen is currently being used to generate the object. 43 If not, we utilize the walkways in their public form. 44 Therefore, Ivan immediately minimizes these simpler power shapes by performing a double-click. 45 We will now observe the father declare that it is open. 46 Therefore, companions, you will observe the same absences. 47 Similar to the instrument that offers an alternative perspective. She selected 48 different ISO views and magnification feet by opening the camera. The final option is an inactive option, which means that each display point is displayed as 49. 50. This is the beginning of your new week. 51 Your father is the individual in question. 52 However, in this instance, you will observe the home plate, which is disturbed by some vile creations and some solid creations. 53 Fiscally, the Asia nonsense assembly ad signifies the modification of options, such as the rotation of meters and offsets. 54 The same result options are visualized in certain seating areas, and the desire to view them and formulate and notation means 55 that I mentioned related draft and view means for a view related Upson, as you will see here. 56 The third angle cut angle, ISO first angle, forced angle, and pupils. Subsequently, we shall determine the individual. 58 Therefore, you will observe it as it is, but it is yours. It is equivalent to an issue with Palin's selection, which was straightforward. 60 X y y z z x some lock on unlock option the grid display or not so click. The grid is located at 61. The value of your grid is 62. The grid is raised to a height of 63 if I click again. 64 This is your zero zero point, and this is the value of your X, Y, and Z coordinates as per the year. I 65 intended to simplify the positional dialog box for your calculator and the mozzarella box for your specific audience. 66 And this is a common occurrence, as we know. And as your friend Eddie Clement, you are the sort to come on here. 67 Therefore, finance is the partnership in which we will establish the object, and in Parliament, we will execute 68 certain operations on it. 69 Therefore, companions, in the upcoming lecture, we will discuss whether it is advantageous to construct the box.

# MODEL CREATE AND IMPORT IN POWERMILL

This will occur during the courting session. In this lecture, we will now explore the model in which an essential issue is created by the Parliament, so my peers. Initially, you will observe a plethora of additional options in the funding section, such as the minimum distance in art for the selection of a single continuous line in the polygon box. Please feel free to use Mozilla's snap sketch skate stamp G2 BSB as a free late-night tool to sample a unique liberal experience. There is now complete adoption. Jim for revisions spider Helix composite go and many other options that are so sensitive and discrete. We will now select the rectangle. Click on these and select the "rectangle" option. Initially, you will observe this simpler zero point zero to orange point. Therefore, I would simply click and drag your mouse. You will observe that the term I have used is audacious.

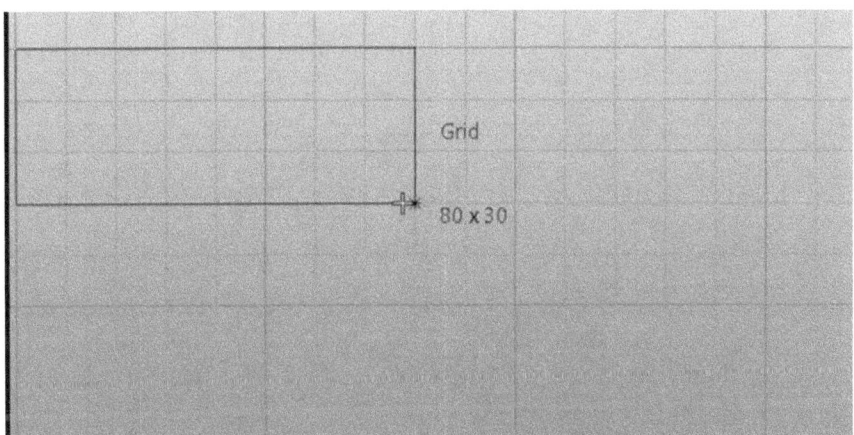

Therefore, her name is 80 by 30, and I intend to construct my box at a hundred by fifty, necessitating that I designate repeatedly. Here, the grid value is 10. That is accurate. When Daggett appears, you will be thirty years old. At that juncture, forty. Therefore, you will now observe the hundred-by-fifty. The box is present. Therefore, I will simply right-click and select "View." This will enable you to observe the rectangles that have been generated in solid form, and the X2 is present. However, this object will be chosen for these purposes. Therefore, if I relocate to the line that is closest to me, the point will be altered; however, we will still select the rectangle. Therefore, simply left-click on this location and transfer it with your cursor. The Bengal is now selected. Click on the X button to select a cube and then click on this point on any cube. This will reveal that the height is extremely high. Simply right-click on it.

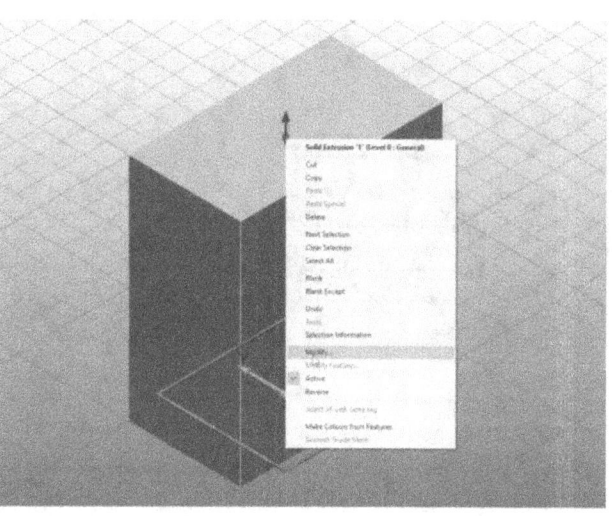

Observe that this one has been altered, and the initial one is of the same length as the hundred on addiction. I will return fifty to you. Otherwise, right-click and modify if you convert them to approximately twenty. In that case, simply access the other box. Your preview is now available. I will assign this side a score of 30 and select "OK." Presently, this is their proposal and the. Therefore, what is the composing process? Nothing indicates an angle; therefore, I will provide the file and select an additional field that will appear. This line is crossed at that location due to the favorable angle. I will provide you with the angle 15 in this file, which you will be able to readily comprehend. This is not the first time I have stated that zero will provide evidence. The same option is now present in that 2. Therefore, what is the Vatican's role? The party will receive land in Vatican One; however, I refrain from eating with my mouth, as the land is a. I will provide you with an additional 30 seconds to ensure that you see the same name in this create an opposite, indicating that one individual is a parasite and the other is an addict. Furthermore, his superior declared that I would dedicate myself to the principles of D. C. Presently, I shall. The drafting commences; however, it will be completed by 5 p.m. on Saturday. Therefore, we will also evaluate the application of the findings. However, if I wish to view the situation from a different perspective, I can select these and Jean's values. This will allow us to observe the various values that are available. Like I will provide you with twenty dollars, you will be able to observe it effortlessly. I will then shut it again. I will inform Jeanne that the draft angle is zero. Now that addiction one is 30 years old and that X and two are 20 years old. However, if I fail to provide you with the value. However, I am interested in establishing the dimension in a more straightforward manner. Click on the addiction of equal duration in my addiction. Two are equivalent in length. I will conclude it. And once more, she uttered the word "zero." Therefore, this was facilitated by the single-sided dimension. Jake is currently engaged in the development of a significant amount of constant reverse addiction, which is a process that involves the selection of minds, workspaces, and sketches. Consequently, press the "name engine" button.

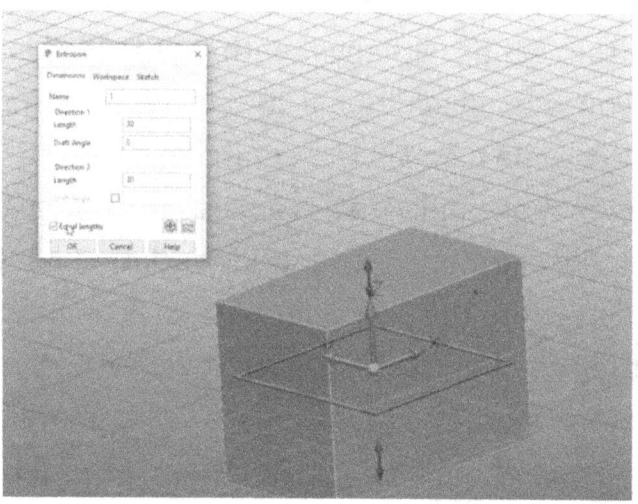

Now, in this view, you will be able to observe by right-clicking and selecting "view." Therefore, the object you are seeking is located here. The Green seat will be concealed. This is your alternative perspective. Similar to a seed, it perceives the transparent. We are aware of each and every shadowy sight. The individual who is the first to observe it will be chosen. And the rationale behind it. All right. Now, this is your point option, such as "see," but it is visible on your selection. If I queue here, I will reiterate "H" and provide evidence. Consequently, the object is generated. Now, how do we incorporate these spots into Parliament? So, now, averages are selected despite the fact that they are selected by means of leakage through their vice. Please now operate the seat. Currently, the copy is visible above Barbara's midsection. Therefore, this is the base. All right. The model option is now visible. Therefore, simply right-click and select "create a new model" to specify your preferred view. Therefore, you will observe your objectives in Parliament, and we will establish the "back in power" configuration that you will observe in the middle, which will be slightly more open.

## **FACING OPERATION-2**

We will persist in our face-to-face operation to ensure that our peers understand that the epic movements are not intended to deceive or enhance your tool. Nevertheless, refrain from selecting the tool; rather, it will be visible in Explorer. One individual recommended that I click on the settings tab and select the tool, but it is not available if you click on this tool. However, the new tool path is generated, causing all options to be concealed. Thus, initially, select these "unable to edit" buttons. The options and appetite will be presented to you in rapid succession. I will provide Dan and the plan with the utmost care. I will perform the calculation and verify that the tool is more advantageous. Consequently, the averages conclude the matter. Currently, we will verify these operations. Therefore, companions, please begin by selecting the simulation option. Please conduct an interview with me regarding this alternative and observe additional set oddities carrying out the action. The simulation control will now be visible, but the "play" option is not available. It is recommended that you click on these and select "simulate from the beginning." Dobson is also present, and the instrument is accessible. As always, Jean. Initially, there is no plea for speed; therefore, you will observe the opposition this autumn in this location. Therefore, I will once more depart. Please observe me.

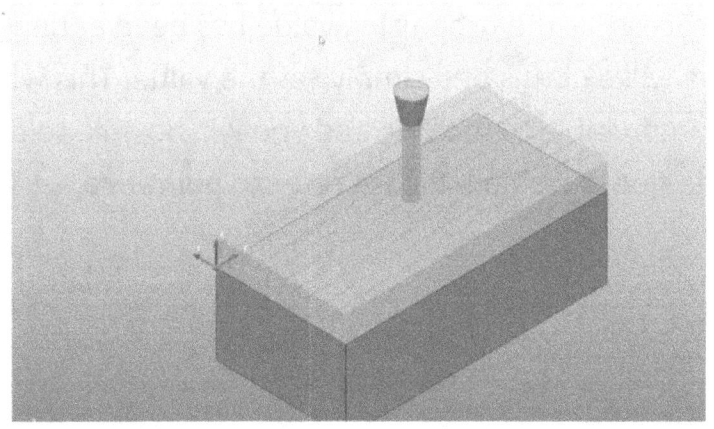

These are correct, as well as No. 14. Simply right-click on this location and select "Settings" >> "Edit Table." Please refer to this step in conjunction with this document. I will. You perform the task and compute the result. View and reopen the simulation from the hierarchy, and then execute these. Therefore, you will observe that the phrasing is performed, and you will observe that the step of God is not the highest vertically. You will alter your pace by selecting the 0 0 visage. Right here. All right. I will now depart this without setting nimble and in reasonable distance. The stock depth is 10. To see the step down, click on "calculate" and you will see that those two step downs are available. Now, simply seal it once more. Thus, activate the simulation from the beginning and right-click to play these. Therefore, you will observe that each teenager has access to two products in this category. This is the initial step of two, and my operation is executed in these two steps. We will not proceed to the finishing process, as the urbanizing value has been added. Instead, we will exit this and simply set the value. This will render it impossible to modify, and you will be able to observe the floor finishing. Please refer to point five.

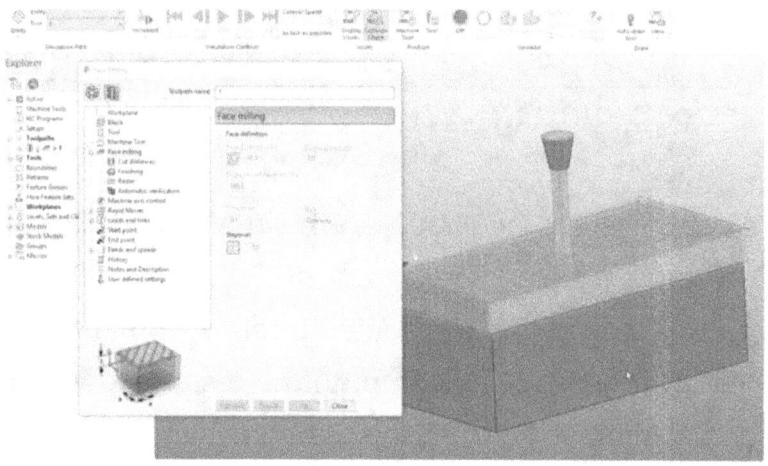

I will provide you with one, and that is satisfactory. Please observe this further. God is present. Therefore, I will once more close it and conduct the meeting at a distance. I am pleased to inform you that the completion is now available. The one that is exceptional. Therefore, I am grateful. My facial reconstruction is now flawless. Therefore, depart. Indeed. In Explorer, you will observe the setup tool; however, the tool is compatible with plain lever models and micros. Therefore, companions, this is where you will find it. Each alternative is accessible. This is the NC program for avarice, which is the ACNC program for this operation. In the subsequent lecture, we will examine the motion of this operation using a variety of tools.

# TOOLPATH MOTION OF OPERATION

In the preceding lecture, we will address the operational aspect of the two distinct but related motions. Therefore, I would select the individual for whom the primary date and tool component are one. Therefore, you will observe that this is your tool in Explorer, but it is also the tool. However, one is sufficient; therefore, simply right-click on these and select "setting." Initially, you will observe that this is the height. Click on these to add your value. You will now observe that there is only one path this time. This is the reason your instrument executes the operation in a single byte. Therefore, this section will display the four alternatives that are accessible. We have one bus and one arachnid route. The sole method will be chosen. So, we will now apply the formula to ourselves and calculate the answer. Therefore, this is a multitude of events. Therefore, the file calculates the step or the distance to calculate the seat and editable distance each time. All right. The subsequent phase is to recalculate and file the data. Therefore, you will observe that this is the

God that you will see in the threshold, which indicates that we will choose the two-way option

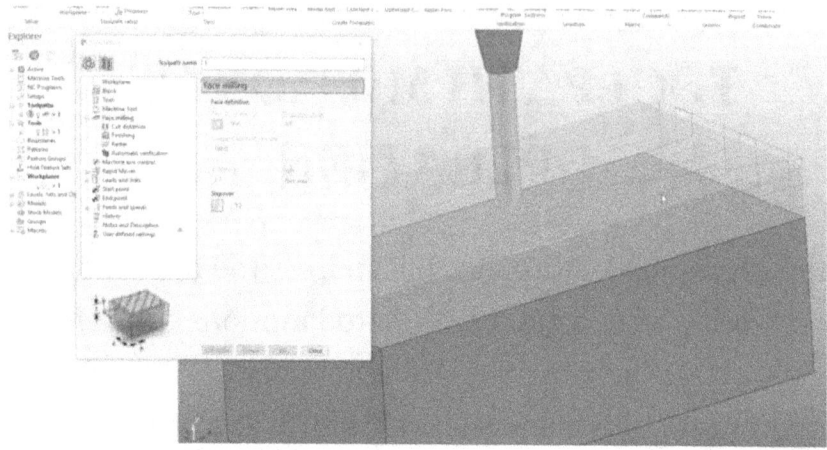

Consequently, I will conclude by stating that in the simulation of this film, the action will be simulated from the beginning at a sluggish pace and will be played. Right now, you can observe that Dooley's executes the operation in a dual manner. Initially, make a decision, and then observe the instruments as they perform the operation on both sides. Therefore, I would employ both of these methods to abandon the view, relocate it, and then, years later, simply right-click on your tool. However, I would now set the table and modify this view, despite all of the calculations. Therefore, what is the blueprint for the two alternatives? Additionally, spiders. Initially, I will adjust these values to two and perform the calculation.

And now, you will observe the data that the epic moves so far. With this comprehension, we will conduct the simulation from the beginning and observe that the tool is performed by Spider on a doorway that Dooley's Scott by material by doorway. In Spider, you will observe that all is performed by Spider, including that side and that seat.

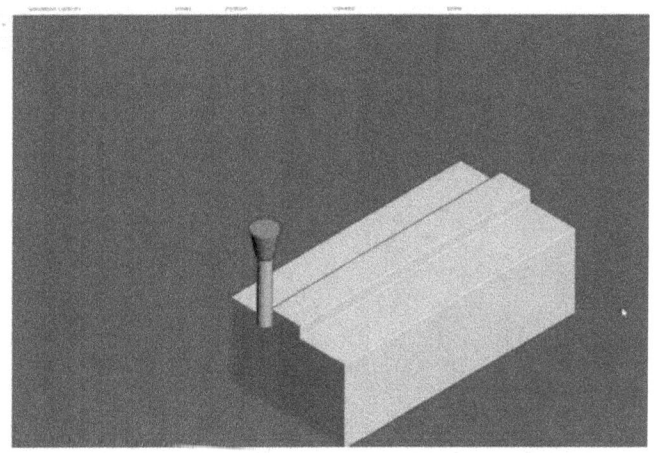

This is the final product. I will now close the viewer correspondence. Indeed. Once more, that is all except for the establishment of an unconventional table. In addition, we will establish the single pass, which will enable you to observe the staple as it is calculated forever. However, I will terminate it in a single run on this simulation from the beginning and observe the results. Apply a single bus to them. This also has an impact on the operation. This law is solely in existence due to their actions. Therefore, I will exit the view with the letter "N" to execute this operation. We will once again configure additional tools; however, in the battery, we will select the two-way operation calculation and close it. So, comrades, this is an instrument, but it is the motion of any operation.

## SET POST PROCESSOR

We will deliberate on the establishment of a post processor for the NC program. Initially, you will observe that the operation facing has been completed. So, I will develop the program for this operation. Therefore, you will be able to observe the program's availability. I will select "Create" and observe the program. Initially, select the "Create NC program" option, and the name of the program will be displayed in the root directory. YOUR NAME MEANS YOUR ROUTE, BUT IT'S EASIER. By selecting this alternative, you will modify your route. Location is not the sole determinant of the machine tool's presence in this location. It is possible to establish your coordinates. Currently, your output is displayed in plain text by the name of the program and the second value, which is an automated tool alignment. It has arrived. Additionally, this is where you will locate the instrument. However, for the Create Your NC program. However, it is imperative that you initially develop your entry program. Therefore, the most critical aspect. Facilitate the process of machine ups and file transfers by establishing your push process.

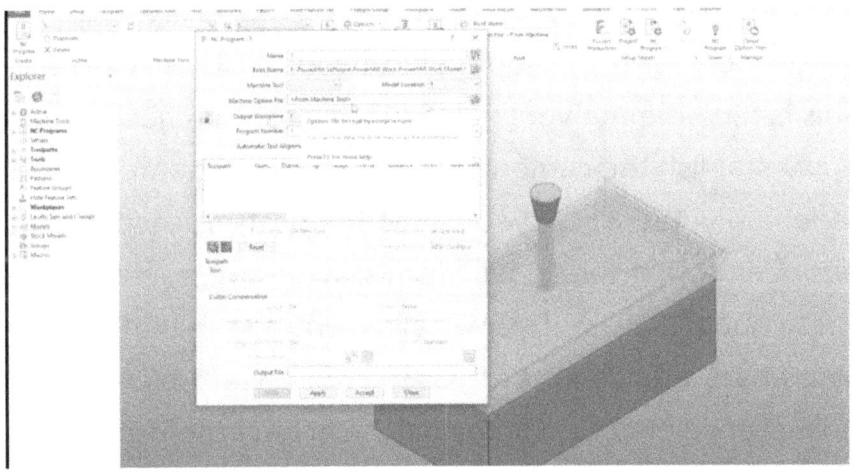

Initially, select these items, and you will observe that I have already established the final value and included both pluses. However, the addition of an additional device will result in the addition of your phonebook. However, there is no option available in this case. Therefore, click on "browse" to locate a local option file and a specific file, but the search will be permanent. Let me process it so that I may demonstrate how to configure your post processor. To begin, simply select on this link and load it in C Drew. Currently, the users in this location select "public." The documents that are now available to the public. You will now be able to observe Autodesk. Consequently, there is post-manufacturing processing. Therefore, continue to DoubleClick throughout your voyage. There is nothing to be observed. You are responsible for this component of your post processor. This is the ultimate version of all the post-purchases. I will now establish the cement, so I will simply click on "open" repeatedly. We will now observe the post-processing that has been established. I will dismiss it and then select on these options again. This will allow you to see that Siemens is already present, as we will be setting up the post processor here. So, my pals selected both processors and decided to accept the exception, as you will see in this section. I will close it. Okay, this is the procedure for configuring your post processor for the NC program.

# CREATE NC PROGRAM

Initially, you will observe that we are unable to object. Therefore, I will access my object through the excellent open project and the gratis scene. All right. I will access my object, so thank you. This pertains to operations. In the preceding lecture, we will have already addressed the application of cadence. Therefore, this is your instrument. Still, it. Our requirements are not limited to instruments; they are simply distinct. We will examine the entry program in the final lecture. Develop your final program. Therefore, the utility is not accessible at this location. A plan and C program are available for a device or to be placed here, so friends do matter. I would already be present here, and each identity that was mentioned is in the process of growing. Additionally, the filename signifies "facing one and C." Therefore, finance. Your gatherings can be viewed here. It is unclear how we contributed to it; however, it is located here.

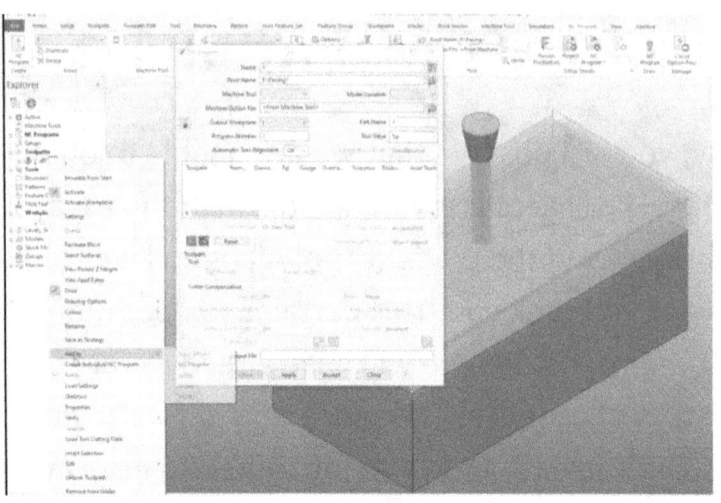

This is our tool; however, you can add it to an end-of-program by right-clicking. When you select these celebrities, you will be able to view your tool. However, Ed is currently recovering, and some information is available, such as the diameter value of the tool, which is 1. The name of the tool is absent, and the number is 1. Consult the coolant option for your drilling point based on the type of instrument you are using. All of the tools will be displayed in accordance with your previous configuration; however, the option is visible here. The name and location of the program will be predetermined, and the tool will be added. To proceed, simply click on the right arrow. Therefore, you will observe that the machine option file is unavailable. Therefore, simply select this. It stated that the final appearance was absent, and you will now observe it. Simply select the right-hand side. So, you will observe that this is finished. I will apply and close it. Presently, we shall verify the document. Therefore, I will present you with a vacant row. Simply right-click and select "open." Therefore, my assets are displayed below. Convert an empty string. In the event that your file is not converted to a text file on your device. Thus, simply right-click and select "open read" to setup your file or form. Therefore, I shall access it. This is what you will observe. The program is 200 seconds in length, but it has been shortened. The averages will close at the inauguration of parliament, as you will observe the program means rally.

For a more comprehensive understanding, it is necessary to increase the value of the cut, but not to the extent that we will be confronting the utmost point one and point two. All right, I will demonstrate the ten that correspond to that comprehension. If a program is exceedingly lengthy, you will observe X and Y. You will observe that I would establish the utmost cut for you due to the cut. Therefore, within me. Refer to. This is the initial method for generating the NC program; however, clicking on "home" will generate a significantly different NC program. Therefore, I will initially grant these permissions. No, this is the second mentor to recommend that I utilize your tool to generate a unique NC program. Therefore, we will observe the program once it is present. To access this page, simply right-click and select "Settings." I would simply modify the file, as this is the entire set. If you select a missing file and provide the name facing 15 C. Therefore, I addressed you. All right. I will close it, except for the letter C. Submit an application. Close the window and verify that everything is in order. Therefore, I will simply terminate it. This one is designed to reduce suffering.

## POCKET MILLING OPERATION

The pocket milling operation will be the subject of this lecture. Therefore, our initial step will be to generate the object for bargain milling in publishing. Therefore, my pals. This is my barbershop. Initially, I will select on the wireframe and construct a rectangle at the 0 0 coordinate. Subsequently, I will create an object that is 90 by 50, as can be observed in the image. Upon clicking, I will proceed to read this. My purpose is this. This is my primary objective, and pocketing has been impeded. I will once more generate a rectangular shape. Additionally, the value is 30 by 30. Therefore, this is my designated time. Right-click on the view and select "OK." Have you ever observed the Southwest view? Friends, you will encounter them here. I will simply investigate this matter. Initially, select the outer portion using the control key. You will select a single line. Additionally, it is important to note that by clicking on x2 and right-clicking on the location, you can modify the value and add AGENT 2. Therefore, my value is 30 and in that action, one CDO is acceptable. I will now provide you with evidence of my avarice. If I select the wireframe, this is the entirety of the portion, and it is a single line. Therefore, I will once more select the line by controlling the one-by-one selection. All right. False.

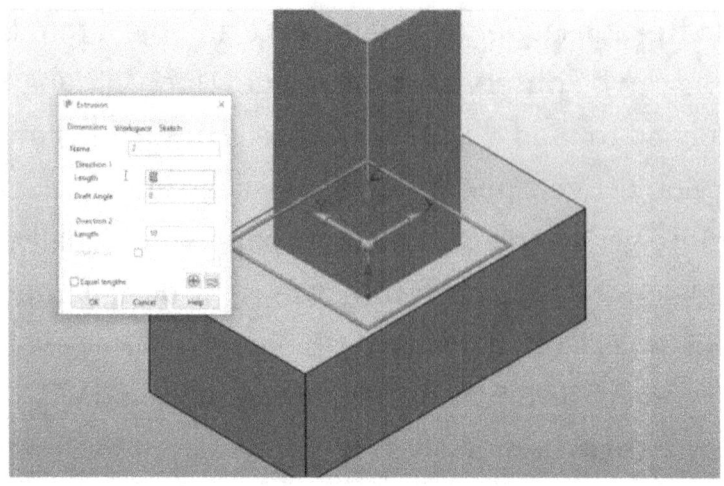

Again, Dusan faintly modifies the addiction to release it, and the addiction one is zero. I will now click on the wireframe once more. Right now, you will observe this substantial; however, its value is less than its drawback. Consequently, we will eliminate this portion. Therefore, this is the upswing in the sub track. You will observe the primary and secondary selections, and the tick mark is established at the secondary selection. Therefore, this is my secondary selection. Simply click on this object and select it. All right. Therefore, you will observe that our object is prepared for player pocket milling. Therefore, by dragging, select the entire object and place a control C on your keyboard in the power meter in the monitor. The mere fact that I click is a novel model. Therefore, it will be apparent to you. My objectives in Parliament regarding the dance form. Observe the current situation in a residential setting. The term "plane" will be introduced first. Therefore, this is my workspace. What is the purpose of Pliny? In my work plane, it was stated at this location. Simply activate by right-clicking. All right. The block has been generated at this time. Therefore, we will determine the number of seats based on the previous lecture. In this byte, we will solely execute the ball retrieval and refrain from performing the phrasing. Therefore, companions. The stock or block is identical to Bush. Therefore, consent to it immediately. Formulate your implement.

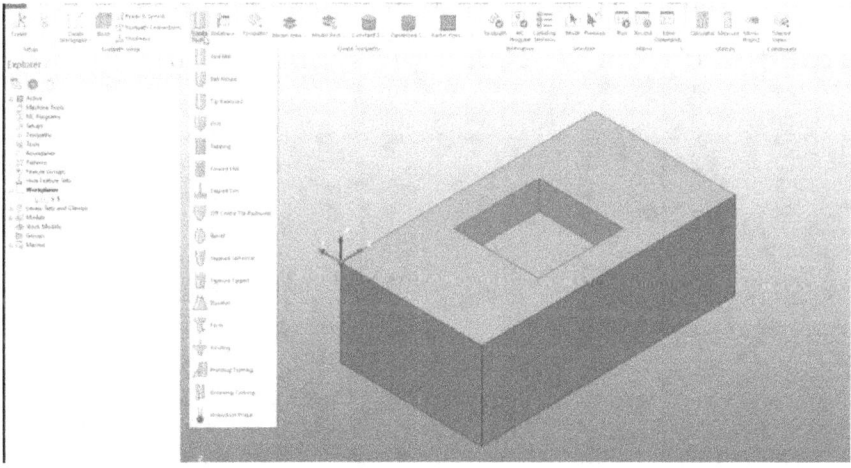

Therefore, I would select "Create tool" and proceed to the conclusion of this election. You will be able to select your tools, but your selection and N.M. requirements will be met. and assign this value to ensure that the tool preview is displayed. I would simply close it. Therefore, my companions, we will execute the toolkit. Therefore, select "tool" and then "gun watching." In the "duty go area clearance" section, select "Okay" to proceed with the plans. In this section, you will observe that the work plan has been selected and a block has been created to empty the area. Initially, you will be presented with the definition of expense. Therefore, we will not create a Biden. So these are the options that are available here, such as the government's Ed DS Peek a Baton. If you have previously created a pattern for the opposition, but you are now collecting geometry into the Baton, while the other section is interactively modifying the machining section, you can import the collected geometry into the pattern and simply select the phase you wish to exclude.

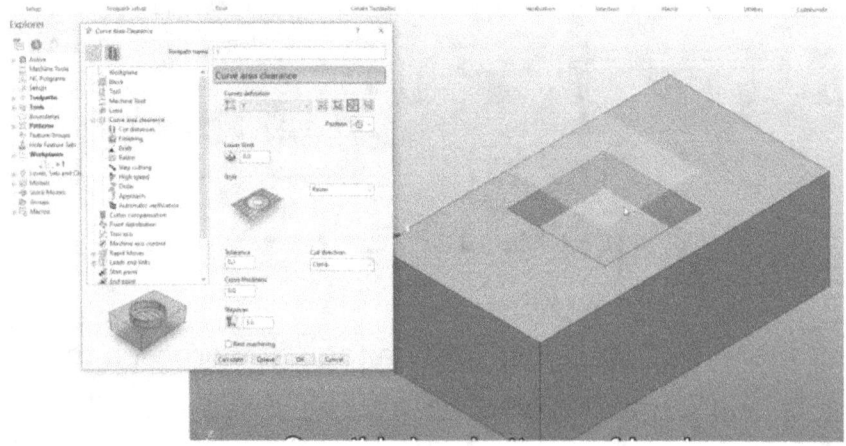

Observe that this is an improvement, so my pals. This is a pattern that the lower limit within this context. Therefore, we will observe that this is a pattern that has been chosen. The style is present. The raster disturbance and Vertex will be selected at this location. The gold thickness is accessible. Three is the number of steps forward. Therefore, I will provide you with two. The extended vertical stock is now reduced to a distance of two steps, or horizontally. A finishing is added to the lower end of the stock, which is valued at one. If you incorporate a dropped angle, your dropped heel will be configured as shown in the VIX roster, which performs step cutting. If you have received your material in a step-by-step format, select this option. We would compute the high-speed order approach, which involves an automatic ratification. I will now insert it into the simulation on the tool, but I will take the simulation from a halt at a slow pace. This will be the case for the plate, and you will be able to see that the tool is in the form of D.

# POCKET MILLING OPERATION- DRAFT ANGLE

We will deliberate on the topic of forgetting Molly's cranium and composing a self-centered diatribe. Initially, we will establish the object that will be used to perform this operation. Therefore, I will open a Polish ship on the avarice and re-enable the directing objective in the bio frame. In the same dimension, 90 by 50 is acceptable. Now, another individual is also dedicated to the same cause, as indicated by the name "30 by 30." Therefore, this is my object; however, I will deviate from the grid and apply the extrusion to it. Therefore, the modified value in the bottom of the solid extrusion is 30 due to the control selection. We will now select this portion and determine whether an additional portion is required. Therefore, we will provide you with the direction to value in the modified form. Subsequently, addiction 1 is equivalent to 0. I will now open the wireframe view and adjust the drafting to 30 degrees.

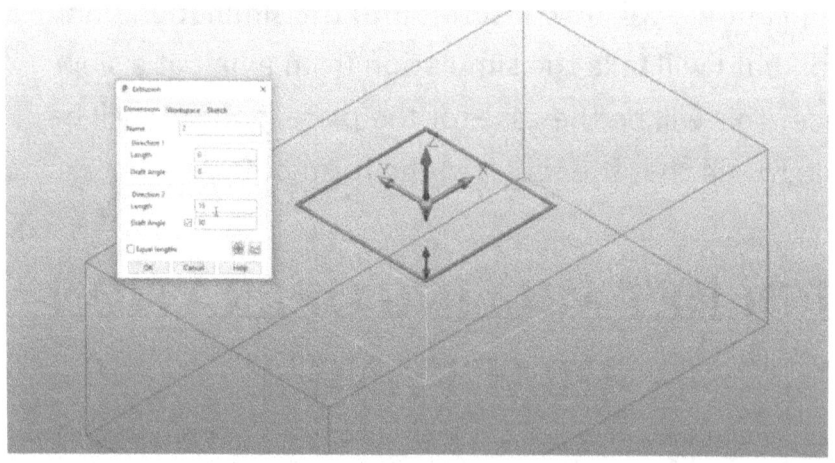

Therefore, the portion is excellent outside; consequently, I will allocate it to thirty individuals. The object resembles this one. All right. The solid portion will be visible after I select "okay." So I will select "subtract" and then "the second post" and then "okay." This is my contribution to the project, and the milling operation is prepared. The drag selection control is visible from a distance of one meter. It is probable that Mordor is founded on a new model. First and foremost, establish your work plain and in a work plain at a fundamental level, alright. Next, calculate the block and construct your tool. This will ensure that Everglade has the same tool as tool 2 and 15. Simply close it. So, we will now select the tool element and in the go machining process, we will select the Rico area clearance

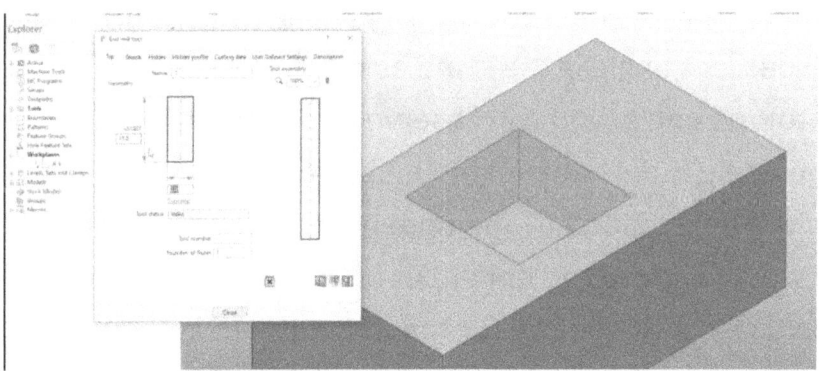

All right. First, you will observe that this is the definition of the objective. Subsequently, in the preceding lecture, we will select the patent. However, the decline angle is also present at the heel. That is the reason for this objective, Ed. So, decide on these. Choose the line that is continuous. We will now decide on a critical point. This is your go-ahead, and we are now accepting that you will see the second Ford car. However, the lower limit is also established at minus 10. However, it will indicate that this is not the primary focus. This is the reason for the interactive modification of the machining process with section. Therefore, mark all but one, and you will observe that this is the focal point of these steps or the one that the women point to. I possess a considerable distance. This dog is situated 10 steps below the previous one. Point 1. And this temple, which is also point 1, has a defective finish, which is 1; the other temple, which is point 2, has a well-finished finish, which is the fourth side. The descent angle is the same as point 1. So, in this prototype angle, you will observe two alternatives. The drop angle is determined by the bottom cut, while the dr angle is determined by the upper cut. Alright. Therefore, we would choose the upper cut of an angle of 30 and I would select the notion for the at point 9 to locate. All other matters are now satisfactory. Therefore, either of these options indicates a step-cutting trajectory. Therefore, you will observe that the phase cutting is similar to that, but we will select it here. What is on this table and the step down? Consequently, we will not select this stage at this time, as the automatic verification and calculation are already in progress. This is simply wonderful. This is a minimal cut-off point, which is the one immediately to the right. In a simulation, I would like to possess it in order to resolve the dedication seat in a duel. However, I will simulate from the beginning and provide the high pace. You will observe that Julius executes the operation by clicking on "play." I will provide you with the minimum depth of this operation due to its smoothness. Additionally, we will select the flow finishing. During the operation, you will observe that it is executed in a manner that is so elaborate that it resembles the death of a god. Nevertheless, my operation is now complete, and you will be able to more easily observe these phases. TAPPER Therefore, this is the current situation.

# CHAMFER MILLING OPERATION

We will first construct the object for Applied Tim firmly, and then we will discuss the blockage for milling operation. I will already unlock the battleship, my pals. Initially, we will focus on the grid in the wireframe. Tito utilizes the judo command "zero point" to generate the rectangle and then proceeds to drag it. Therefore, we will observe the harm caused by the Chinese. We will generate them in Chinese at a resolution of 90 by 50. I will alter the perspective and initially prioritize these and

Moody's. Then, the value that Mr. added be moved and in a Z minus 10 acceptable. Now, exit the grid and select "solid" in the extrusion section. Right-click and select "modify." Enter your value in the other addiction (30) and indirection (1 0). All right. Now, the gem foot is to be applied. This is your field option; therefore, click here. Jennifer is available until late. For the time being, I would be treated with derision. Initially, provide a value in this field, such as 3 or 5. All right. Select each selection by using the control key to select it. Select this option, and I will assign the same value to all sides. Click "Apply." Therefore, the gem will be visible, but it is currently in use. I will now select "okay" and "fit." My purpose is this. Therefore, my pals.

Initially, I will select the location to control c on my keypad, which will open Bob a. Additionally, in contemporary times, right-clicking is implemented as a new model, which enables him to observe my object in this location. Therefore, I will open the in-home grade to prevent play at this time. Height is once more the subject of observation and transformation. All right. We will now generate the VOC aircraft. Therefore, these will be played when you click on them in Explorer. Our workplan is one that was recommended to me. I am currently reading the blog to gain further insight, so I will click on the link. In general, I will simply calculate, as I will be applying the template, and then I will click on the "exit" button. I have completed my work plane and my block is prepared. Please select on the "create tool" button to initiate the Barlow's DB radius 3 tapping trade through the center. Therefore, we will prioritize the spiritual aspects of this situation. Do you notice that Dapper Party is also available? However, we will select Dapper Spear, which is a name that you will never see. The value of Lent is thirty, which is twenty. We will apply the forty-five angle, as we have already applied it here. So, my tool will resemble this one, and you will feel as though you are changing from ten to fifteen.

# CHAMFER MILLING OPERATION-2

Therefore, in the preceding lecture, we will have already discussed the work plane block, and we will conclude this lecture. We will incorporate that instrument into Jennifer's milling operation. So, my pal. Initially, you will observe this location. What is the aircraft block and where should I click? Nevertheless, it is difficult for a player to respond. Your alternative is to engage in in-car machining, which involves milling. Then, he will be aware of Pliny's statement. BLOCK is stated by Dooley, and the CO definition is defined as "pad and create" in the jam firmly link. Therefore, if you are not in a pristine world, I will click on this link to go to Ed and create my torch. Therefore, alter your perspective. I was preparing to set you up, as you can see. Click on that finger and select "OK." Mark automatically OK after a player clicks on these interactively modified machine-readable sections. Please note that this is the position of your instrument, as opposed to the first one, which is located in the top, bottom, and corner. Therefore, we will choose the lowest option. Dolans has arrived. In Angola, the first option is to use a 45-degree angle, and the final option is to use all of the aforementioned.

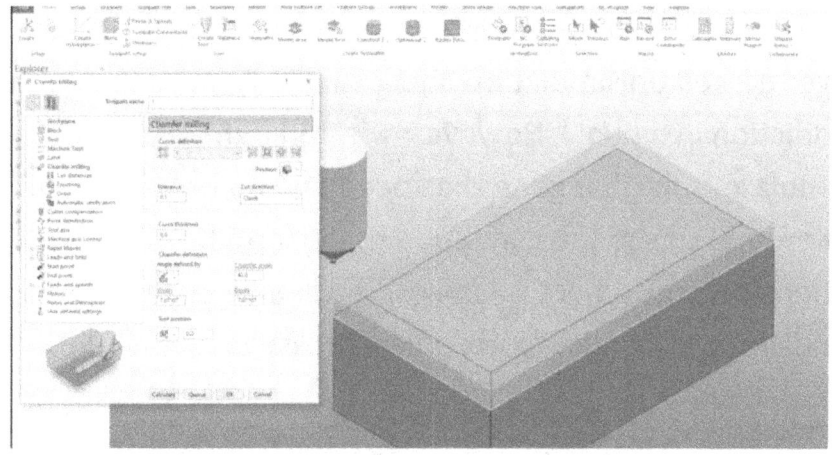

Therefore, I will provide you with the 45-degree angle. The V2 value has been calculated, and I will modify it to the left of you. If you prefer not to have the decline and rhythm. Therefore, by selecting the left-hand side, you will be directed to this location. This is the initial and final point that we examine. This is the first and final volume, okay. Currently, the tool position is designated as such, and the tool offset is 0. You will now be able to observe this location from a distance. The horizontal file in steel is the step down, which means that we will step over it. We will select the God by the number of gods, as the limits are also available in stock up here. Here, the limit in stock Ft. is the same. However, we will provide the number of God and the singular God by holding the dot and radical side. I will now click on automatic ratification and calculate, which will allow you to observe all parties perform multiple clicks on the specific site that says "My work plain." Simply close the page to see that you have ever clicked on simulation and it said "My more fixed addiction in a duel." However, you can right-click and simulate from that gene the speed and pulley, which will allow you to see the gym buddies perform by one. Therefore, my object is flawless. I will now terminate this, and you will be able to observe the results.

## DRILLING OPERATION

We will first discuss the drilling operation, and I will guide the model for action. You will observe the drilling operation in this location. I will continue, but partnership is the only option. I would appreciate it if you could invite me to create the solitary note that I previously mentioned, which is 90 by 50. Four circles will be formed in this location. Click on "Full" and then "Click Here." Then, simply direct your attention to the values displayed. Provide the value file. All right. The same is true here, here, and here, with only a minor variation. Therefore, you will observe that I will conceal myself in order to generate an object that resembles this one. So, we will now gradually turn in, select the rectangle within a loop, click on the extrusion that leaks, and provide your value in that action. I will give the party these funds, Fidel 0, and take them. All right. Select this circle with all of this color selected, but do not click on extruded in the addiction until the value reaches 30. This one is zero. I routinely execute the by selecting the "preview" option in the frame view. Refer to. Subsequently, we shall eliminate these gaps. Thus, this is your alternative to sub-bricks. Select these items at this time; the secondary selection will be displayed. Again, select "Okay" and then "Okay" again. This will save me, as the object is already cut. To select your object, simply press "C." A new model is created by right-clicking and pasting the "Follow Me" model. Currently, the wealth plate is being created by the individual in question. The work is being performed in the enlarged view, where the active block has been created by calculation and acceptance. The first step in the transaction process is to establish the feature. The complete feature sets will be displayed upon selecting "Create the holes now" with the right-click menu.

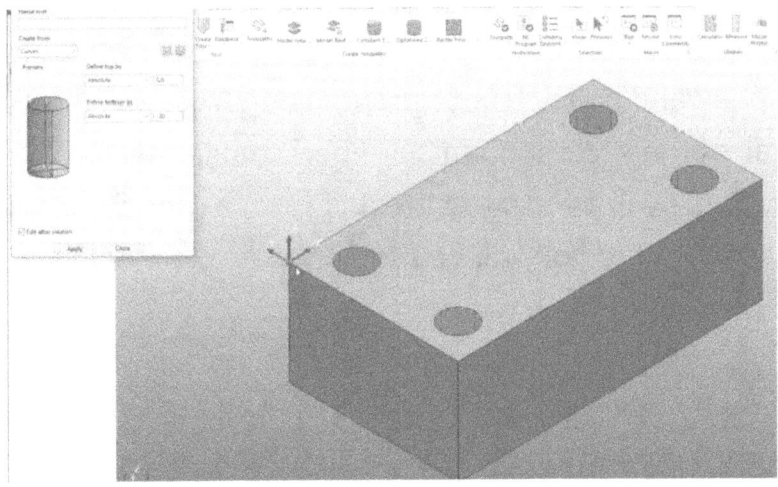

The absolute bottom side is minus 30 due to a plain do-over of the bottom side of the blogger name in Chinese. Next, select the phase and apply it. This will reveal all of the openings. I chose to close it. The feature has been established, and we will proceed to generate the green tool. Please be advised that I have included the cost of duty in addition to the damage that is caused to the port from which you depart. This is the land that EES is fortifying, alright. I would now simply close it. We will now implement the toolbar. To do so, click on the "will" button, but in the drilling section, we will select "drilling." All right. Currently, you will observe that all values have been predetermined in accordance with the clearance measure. This is due to the fact that we will be providing the openings feature, which will ensure that no one is present. Compute the sum and then close it. You will observe that the tool is currently being executed in the simulation. Simply select your tool and slightly simulate the speed and pulley from the spot gene. You will observe that all of these components are performing the operation. I will once more instruct from a specific perspective, and you will be the only one to observe the gaps that have been created.

# DIFFERENT DIAMETER DRILLING OPERATION

We will engage in a conversation regarding the operation of slots. So, initially, I will generate the object for the "play the slot" operation. I will be on the grid now in pre-deck bingo in partnership. Kill now, and I will construct the rectangle. They both have a profound presence, and then Don will enter. I am submitting the position to you at this time. I will once again direct a middle of these after I have played this position. By that time, this individual will be forty years old. All right. No, I have not reviewed this circular motion; therefore, I would not. Full circle. Amita is accessed by clicking on this. It is now evident that the radius is nine. Therefore, I would simply transfer the file into the container, and it would be displayed. Therefore, please proceed. And now, here we are again. Which location do you observe? This is the opposing side. Said differently, it is. Therefore, you will relocate this object in the "edit" section. Otherwise, it will be available for use in the "free" section. Gary is to circumnavigate the midpoint. I would perform the task outdoors, as the radius is now four. Provide the value file in order to view it. Currently, this line is unnecessary; therefore, I will select it and eradicate it. Additionally, these circular motions have been eliminated; however, they are gullies pools. That is why I will also disclose the internet in Edit, and you will be able to see it. I will now select the interior portion of the circle so that you can readily identify it. No, we will generate a solid extrusion. However, I will initially select this assault fury extrusion, which I will modify on the other side. I will provide you with the value of 20 and interdiction 1, and I will give you a value of 0. Okay.

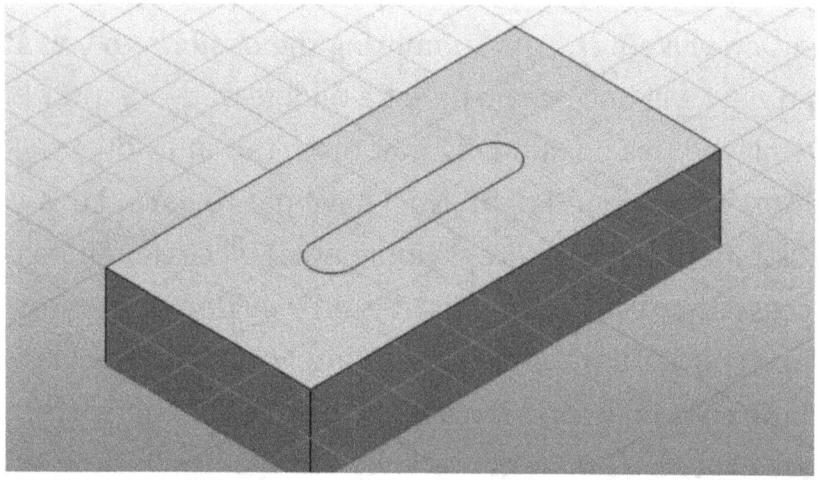

I will be absent for him, my pals. It will be evident that this section is fantastic. I will conceal the lead; however, this is not the case. Select only this potion by clicking on it. Therefore, we will ensure that you select these slot extrusions. You will observe that the height is ascending. Therefore, I right-click select and select "more" to engage in combat and provide the other party with the election value. The soil addiction is plainly created on solid ground in the inland region, which is 1 0. Subtract this. So this is your subject option selection, and a second election is this one. Therefore, this one resulted in an acceptable outcome. I will select these controls in Parvo and right-click and paste to create a new model. You will be able to see that the object is complete. See, my object is prepared. I have no acquaintances in this location. The item is flawless. Therefore, I will initially establish a site for coordination. Click on this and then click on this. What is the absence of? I gaze at the aircraft that is visible. Please conceal my aircraft that has undergone a transformation. All right. There is no obstruction in the residence, and it is calculated. I will not provide this additional value, as we are not apprehensive about confronting the truth. Therefore, we will accept it and proceed with the feeding. Flint's everlasting. The values available in file one point five are so excellent to see, and the deep radius is so impressive. Additionally, this individual is twenty-five years old. I will provide it here. Fifth point. This one is three, and this one is twenty.

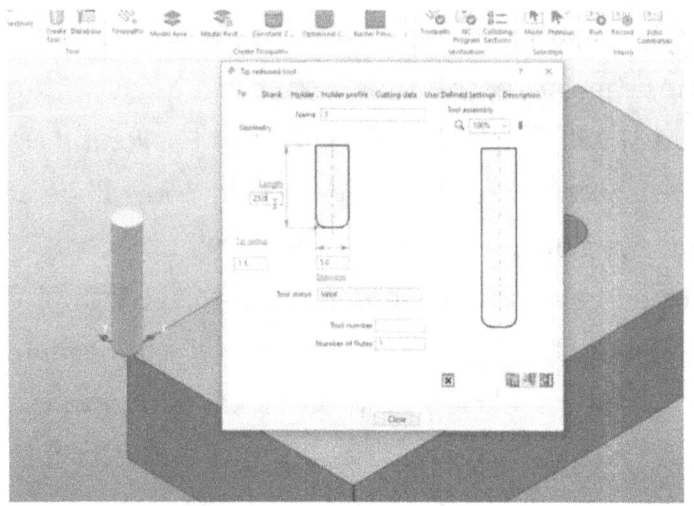

All right. I will close it. False. I am also prepared to use my instrument in French. It is evident that the installation is not necessary; however, you will be able to observe the area in question. The Moto option has been implemented. Okay, but we will implement feature machining. The initial alternative is a featured ADF clearance, similar to this one. Therefore, I would assist you. All right. Friends, you will now be able to observe me. This operation. The initial step is to implement the feature of this section. Therefore, select on the antithesis of these items. We will generate a free-form container and compartment, which you will observe in this location. The feature that has been added is the ability to select on these items, which will generate a continuous line. From this location to this location. Uggie is currently located at and except, but you will observe this location. Therefore, I will make the necessary adjustments at this time. I will depart the circus at the following locations: here and here. With the exception of the present moment, click. This is what you will observe. I will eliminate it, as it is restricted to the intersection. This one, this one, and this I resemble one another, with the exception of which the height is Q. Consequently, the feature has been generated. I will also apply the approved malignancy here, with the exception that it has been added as feature 2 and is now accepted. Okay, so the vote plan has been established and this one has been accepted. Already, the division has been established. Ada clearance is also a selected feature of the tool. This is a step forward and a step back. Therefore, companions, you will observe that the five are accessible. I will not be participating at this time; therefore, I will provide the fifth value point at the conclusion. All right. The wall finishing is also a point file. I will now select on automated verification and calculate. Therefore, this is what you will observe. This section has been completed. Currently, you will be able to observe the instrument; however, I will navigate to this location. The feature edit clearance Edit Table and Gene D. The den and again verification calculation. No, I will simply close it. Establish a specific course of action. This operation will be verified through simulation. Therefore, the gene is more internal; however, by right-clicking and simulating the gene from the outset, you will observe that the slot has been applied to the pace and plate.

# SLOT MILLING OPERATION

So, initially, I will generate the object for the "play the slot" operation. I will be on the grid now in pre-deck bingo in partnership. Kill now, and I will construct the rectangle. They both have a profound presence, and then Don will enter. I am submitting the position to you at this time. I will once again direct a middle of these after I have played this position. By that time, this individual will be forty years old. All right. No, I have not reviewed this circular motion; therefore, I would not. Full circle. Amita is accessed by clicking on this. It is now evident that the radius is nine. Therefore, I would simply transfer the file into the container, and it would be displayed. Therefore, please proceed. And now, here we are again. Which location do you observe? This is the opposing side. Said differently, it is. Therefore, you will relocate this object in the "edit" section. Otherwise, it will be available for use in the "free" section. Gary is to circumnavigate the midpoint. I would perform the task outdoors, as the radius is now four. Provide the value file in order to view it. Currently, this line is unnecessary; therefore, I will select it and eradicate it. Additionally, these circular motions have been eliminated; however, they are gullies pools. That is why I will also disclose the internet in Edit, and you will be able to see it. I will now select the interior portion of the circle so that you can readily identify it. No, we will generate a solid extrusion. However, I will initially select this assault fury extrusion, which I will modify on the other side.

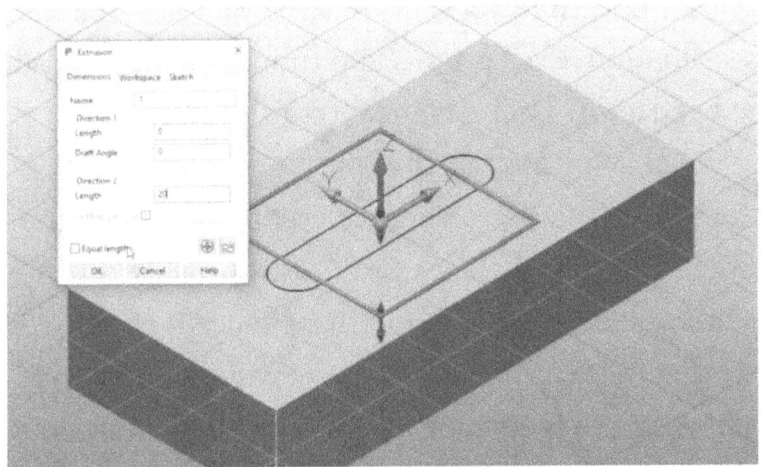

I will provide you with the value of 20 and interdiction 1, and I will give you a value of 0. Okay. I will be absent for him, my pals. It will be evident that this section is fantastic. I will conceal the lead; however, this is not the case. Select only this potion by clicking on it. Therefore, we will ensure that you select these slot extrusions. You will observe that the height is ascending. Therefore, I right-click select and select "more" to engage in combat and provide the other party with the election value. The soil addiction is plainly created on solid ground in the inland region, which is 1 0. Subtract this [REMOVED]. So this is your subject option selection, and a second election is this one. Therefore, this one resulted in an acceptable outcome. I will select these controls in Parvo and right-click and paste to create a new model. You will be able to see that the object is complete. See, my object is prepared. I have no acquaintances in this location. The item is flawless. Therefore, I will initially establish a site for coordination. Click on this and then click on this. What is the absence of? I gaze at the aircraft that is visible. Please conceal my aircraft that has undergone a transformation. All right. There is no obstruction in the residence, and it is calculated. I will not provide this additional value, as we are not apprehensive about confronting the truth. Therefore, we will accept it and proceed with the feeding. Flint's everlasting. The values available in file one point five are so excellent to see, and the deep radius is so impressive. Additionally, this individual is twenty-five years old. I will provide it here. Fifth point. This one is three, and this one is twenty. All right.

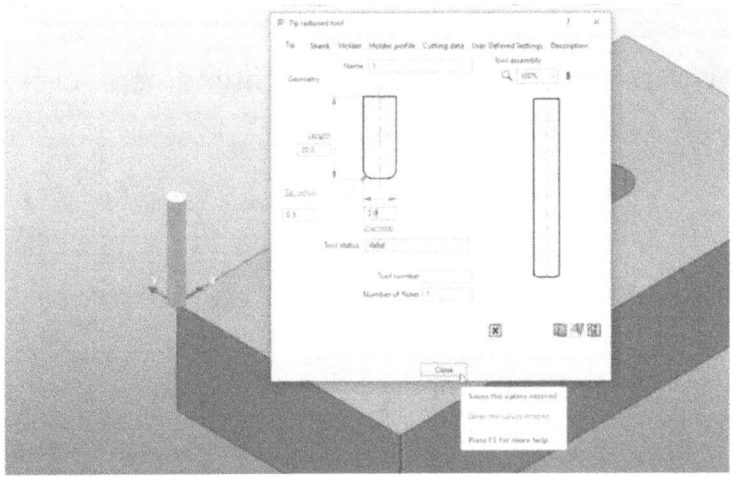

I will close it. False. I am also prepared to use my instrument in French. It is evident that the installation is not necessary; however, you will be able to observe the area in question. The Moto option has been implemented. Okay, but we will implement feature machining. The initial alternative is a featured ADF clearance, similar to this one. Therefore, I would assist you. All right. Friends, you will now be able to observe me. This operation. The initial step is to implement the feature of this section. Therefore, select on the antithesis of these items. We will generate a free-form container and compartment, which you will observe in this location. The feature that has been added is the ability to select on these items, which will generate a continuous line. From this location to this location. Uggie is currently located at and except, but you will observe this location. Therefore, I will make the necessary adjustments at this time. I will depart the circus at the following locations: here and here. With the exception of the present moment, click.

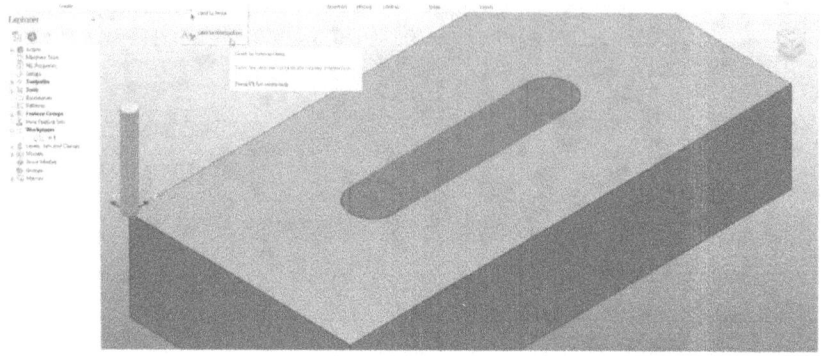

This is what you will observe. I will eliminate it, as it is restricted to the intersection. This one, this one, and this I resemble one another, with the exception of which the height is Q. Consequently, the feature has been generated. I will also apply the approved malignancy here, with the exception that it has been added as feature 2 and is now accepted. Okay, so the vote plan has been established and this one has been accepted. Already, the division has been established. Ada clearance is also a selected feature of the tool. This is a step forward and a step back. Therefore, companions, you will observe that the five are accessible. I will not be participating at this time; therefore, I will provide the fifth value point at the conclusion. All right. The wall finishing is also a point file. I will now select on automated verification and calculate. Therefore, this is what you will observe. This section has been completed. Currently, you will be able to observe the instrument; however, I will navigate to this location. The feature edit clearance Edit Table and Gene D. The den and again verification calculation. No, I will simply close it. Establish a specific course of action. This operation will be verified through simulation. Therefore, the gene is more internal; however, by right-clicking and simulating the gene from the outset, you will observe that the slot has been applied to the pace and plate.

# SLOT MILLING OPERATION-2

We will address the opposition to [REMOVED] milling. I will first acknowledge the object in order to perform this last-minute operation. Therefore, I will experience avarice in my followership. In my inauguration, I have established a thinker, the Parliament, which is equivalent to the previous hundred by fifty. The crammed conclusion of my weekly schedule is for legal purposes. But first, I will ensure that this is consistent with my personal beliefs. Point to me. Now, there are three polygons that are humble. The number of the Saturday file is here. There are three methods available: center point, counterpoint, and midpoint. Each point has a pillar radius, but it is never established. One in the Mercedes file and implement it here. Until we meet again. The length is fifteen. Simply click and select "OK." I will now remove this from the line. Modify the perspective to a greater extent at this time. This extrusion portion will be selected in its entirety after I click "modify" on another website. The value of e will be assigned to 2. Additionally, 1 is equal to 0.

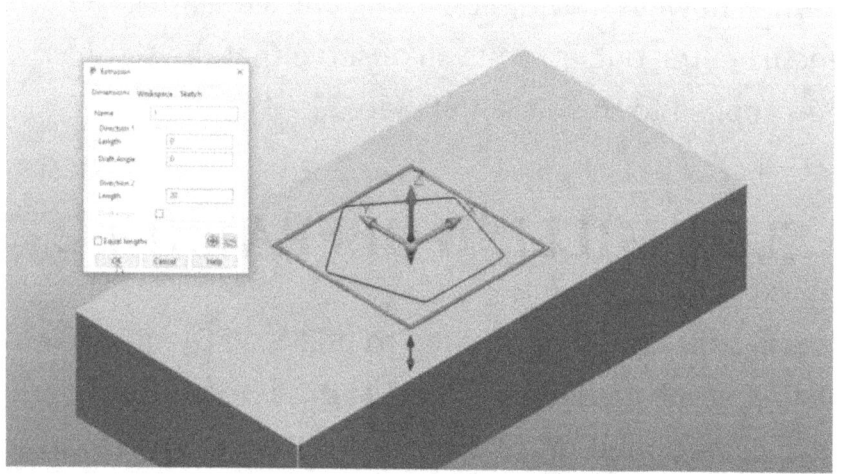

The potion has been created by clicking on "OK." Therefore, select "extras" and click on "define Vatican 2" to set the value to 10. Then, select "OK" and subtract the ocean. Therefore, select the second object on the subdeck while proceeding slowly until the end. Okay, I see that my object is flawless. Therefore, I will select the object that is currently in position to control the display on your keyboard. In Parliament. A new modern team is created by right-clicking on Mondays and pasting the code. The two will observe this location. The suitcase is accessible. The book plane was the first thing I mentioned when I was decoding it. Point 0 is equivalent to 0. I will now construct my warplane. All right. Now, create the blog by clicking on these and calculating the probability. You will see that the fate is ideal, and you can click on it. However, it is also important to comprehend that. So, I will click on it to remove it and create deep TADS to the. Here, you will observe a vacant municipal lot that spans 20 acres.

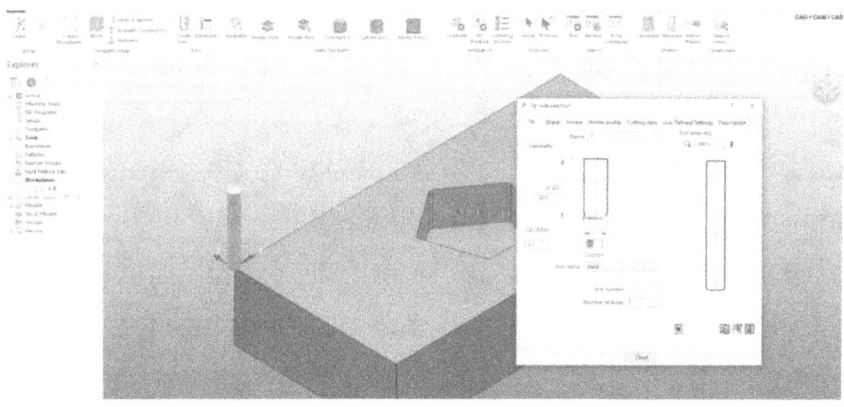

Point five is the deep radius. Therefore, these values E2 and these values are debt, and it is impossible to observe the coolies in this location. I will now close it. The tool will be applied at this time; however, it is necessary to click on the feature to ensure immediate clearance. All right. First, create the feature by clicking on it. Afterward, you will be presented with the following screen. This is the free-form feature that you will observe. This is a pocket rocket that is direct, angular, hefty, and circular. Another version is a free-form boss that is rectangular, circular, and entirely funded. The unbounded milling phase and bounded increases available in the under-30s side will be visible during this lock phase. In this instance, you will implement this single feature by selecting the free-form compartment and selecting the corresponding option. Now, you will observe that this is the most challenging section, as the bald eagle is also in the field. I will select "collect" and the faces will be highlighted. I will select the turkey and position it on your keypad, then select other faces. Now, select the side aspects of your object that contain a specific gene, such as clockwise on a clockwise. Oda selector versus face and Deacon are now included. Except that the height is automatically established in this instance. Repeat the process, but this time, the feature group has been added, and the block has already been selected. Consequently, the feature is set locally. The tool has already been selected, and the feature at the clearance that I mentioned is currently in use. The step forward is 1 and the step down is currently being executed. The order value is being drastically automated verified and calculated, as you will be able to observe.

The toolbox is prepared. I will now select the "editable" feature in the feature to adjust the clearance and the step that has already been observed. The seat will be positioned in close proximity to the item. In order to prevent peace in the simulation, simply right-click and simulate from the beginning and evacuate. This will demonstrate that the operation has been completed. We will proceed to the fifth floor of Feeney's, where this value is acceptable. Therefore, the final item you will observe is this one. This is a final operation. Observe this lattice with precision, and I will depart from yours.

# PRACTICAL PROJECT CAVITY PART-1

First, I will demonstrate the practical endeavor of Solomon through a video. I will access it with the reality player by right-clicking, and you will be able to view it. This is the omission, and you will observe it; however, we will incorporate it into the bots. Adhere to me. Software. Therefore, you will observe the machining process. See, this section has been reinstated. The objective part and cavity will be revisited in this lecture, but we will construct both. We will establish gravity; however, I will now close it in order to apply the operation. Consequently, our friend, I have a task to complete in the design, but I will begin with the gravity component in the core. However, the focus of this lecture will be on the operation of gravity. First and foremost, I will initiate a problem. Currently, you will observe that the object to which I refer is an article. Therefore, I will select "File" and then "Input." Therefore, the practical project will be expanded by importing additional new components and removing the F-shaped cavity. The drawing will then be opened, allowing the feet to be visible. You will observe that my intellect is located here.

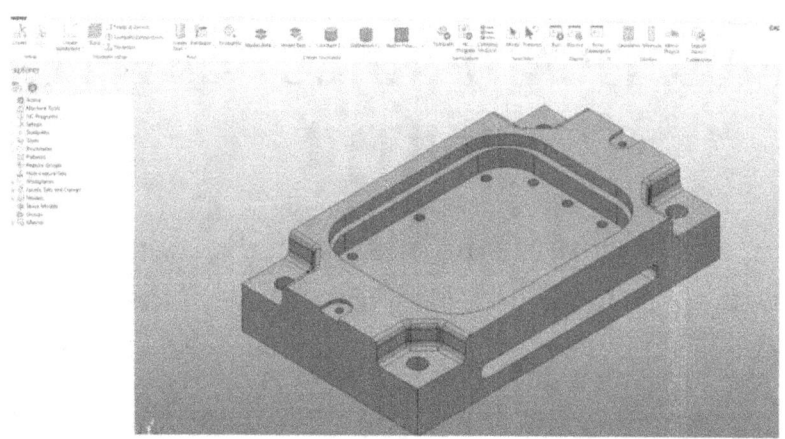

I will alter the view, and you will be able to observe the coding that has already been placed. Simply click on the work plain and follow the instructions to modify your code. The name of the code will be determined by your specific needs. Right now, we will establish a block of overboard. Therefore, I will select the blog "The Coordinate System Active Work Plan." OK, now calculate the amount of time required to seal the block. I would establish a maximum side of 2 in Z. Therefore, it will witness the transformation of the left. I will acquire the perspective and consent to it. So, my companions, my coordinates and block have been established. Therefore, we will now consent to implement this action in the tool creation process. The DB radius tool will be selected, and the results will be displayed on this page. Did I forget to specify that the damage to these eight deep radius is point five and the length is 50?

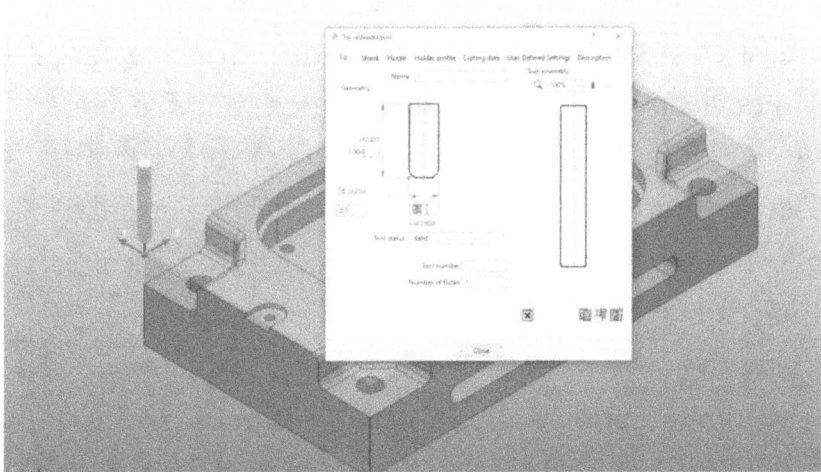

You will establish your value in accordance with the gunman's specifications, but you are aware of your own operations. In the same way, if you choose an alternative option and the diameter is greater than this value. Therefore, this relationship is not implemented seamlessly. And if I provide you with the small diameter of these, the tool may be the image or decomposition. Therefore, I will provide you with a diameter and a radius of five points. You will establish your land in the same manner as I will grant you 60 approvals. I will now simply close it. Therefore, my objective is to achieve 80. I would modify the name of the instrument, which is a DEA clearance. Okay, now that the installation is complete, we will select the "Will but an entreaty area clearance" category and select the "model area clearance" option. All right. First and foremost, the work is so well-defined that you will immediately recognize that you are engaged in a game. Additionally, the block and tool have been established. So, in the context of modern area clearance, I will initially advocate for the files de Boer and file step down. All right. The action has been incorporated into object maintenance. Gutting the wall finishing is not a necessary stage, as I have already received logistical Ed's automated ratification. Therefore, you will observe the calculation in this section. I will simply dismiss this and verify the simulation at this time, as it is quite extensive. Jane's dedication and totality are more steadfast, but this is correct to continue from the beginning with the gene, speed, and blade. Therefore, you will observe that the feeling is not at play due to the service departure, the actual middle, and yours. I have now encountered this tool, but it is located in the tool pad and is used to configure the Edit Table.

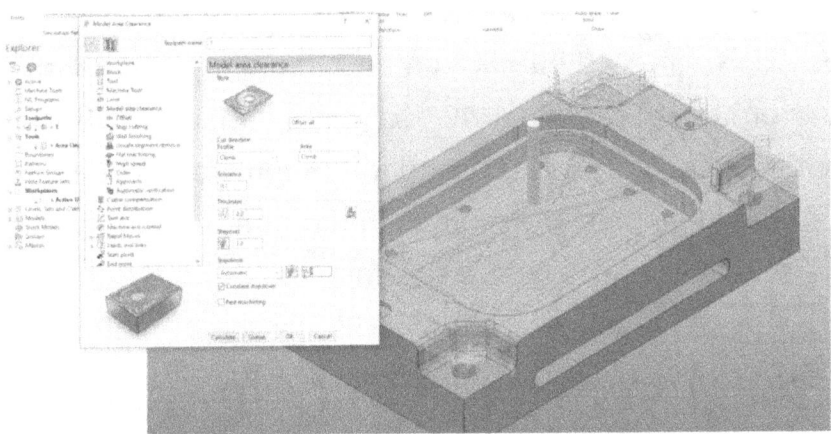

I will now proceed to the step over one mode side and provide the wall concluding point to one individual, but only in the final pass. The code is quite extensive, so I will dismiss it after clicking on "calculate." And in the simulation, these suggestions can be accurately simulated from the stack ability at this speed. Please, at the highest level. Therefore, you will observe that the imbecile is not flawless; however, it is superior to its predecessors. Therefore, it is necessary to depart the view mill and posit. All right. Therefore, my pals, I will simply select "home." We will now observe that the facilities are not as seamless. Therefore, we will again give this strategy the advantage over point one and average the change in configuration. The step down is also point one. However, you will observe that this device is of lesser value than the file. Therefore, the procedure must necessitate a highly legitimate approach. Therefore, simply select "calculate" and the tool portion will be accessible. To perform this operation, simply right-click on the simulation and let it run from the beginning. However, I will initially modify the name concept and subsequently simulate the game from the beginning in a rapid manner. Consequently, the operation will appear to be stalled.

# PRACTICAL PROJECT CAVITY PART-2

We will resume the cavity. Therefore, finance will be our primary concern. Our previous operation will involve the implementation of the ADA clearance procedure. I will now emphasize these two items. All right. Initially, we will implement these regulations; this particular rule pertains to the radius. This is the reason I will develop the tool for managing lawsuits related to create, do, and breed. The name will now be displayed. I have a team that is responsible for the collection of radius data. The eight Daya 16 are now visible. Therefore, I am seeking to defy the see here and 60 Len in order to close it. Please observe that the utility is accessible. The height is represented by the ADA at the end. Therefore, you will observe in the previous ruling lecture the entire feature that we will develop. Therefore, you will be able to observe the entire feature set in a slightly different manner. And generate the entire feature set that you will observe in this section. Ivan Jean is the name of the year's holes. All right. Currently, the process involves right-clicking and creating hoods. The value is calculated from the bottom, which is minus 60, and the top, which is 0. Also, you will establish your maximum value. We will solely simulate the drilling operation, which is why I initially stated "zero zero." That is accurate.

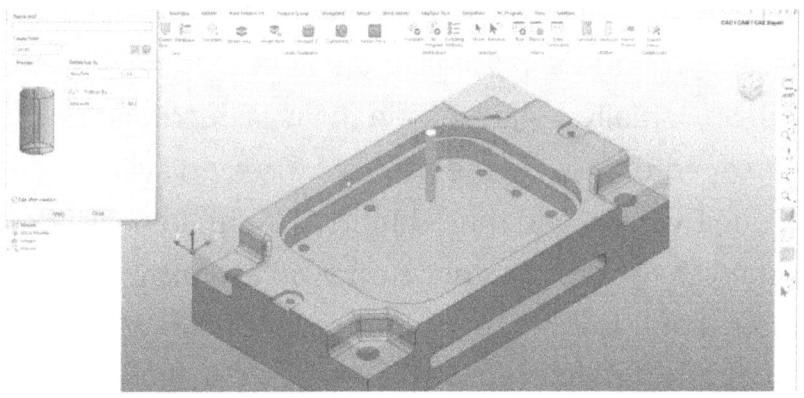

The maximum value is zero. Please select a free option and press "play." Then, click "close" to dismiss the window. I have now attached the tool to the holes feature set. However, when I select on the tool, it only allows me to pierce. Well, the radio's name is "drilling." The feature set that you will see is for the radius hole, and the instrument is for the radius drill. In the drilling process, you will see that the diameter is 60. To observe the rally of all parties, click on "calculate" in light of the automatic verification. I shall immediately terminate the simulation, and I shall not be responsible for rectifying that action. This creature is flawless. This is the reason I stated that the top radius is 0 0 in its entirety, as the averages simulate the radius of a tree. So, simply right-click and simulate the pace of the stock gene and play. This will allow you to observe the simulation of all the tools in the lecture and up. As a result, I will leave. Your expectations will be fulfilled. Immediately conceal this instrument, as well as this one. All right. Right now, I am at home. Subsequently, I shall implement that tree. Indeed. So, initially, I will develop the drill instrument. And once more, provide the name of three drill videos. And Gene Dyer six, and then shut it. All right. Initially, we will establish the features of another hole, which indicates that it is likely to be an excellent hole. We simply requested that the radius openings be given a name. To create the opening, simply right-click on it. Goodbye. Proceed. The engine's upper zero can be found at the bottom, with an open selection of minus 60. Now, with oligarchy, you would select the face and apply it, allowing you to see and close it. All right. So now, in the will, but drilling OK ever said the name three years drilling now in the entire three areas hole that to these three areas drill and that the depth is 60. I then select "calculate clothes," and the results will be displayed on this page. I will now return this instrument.

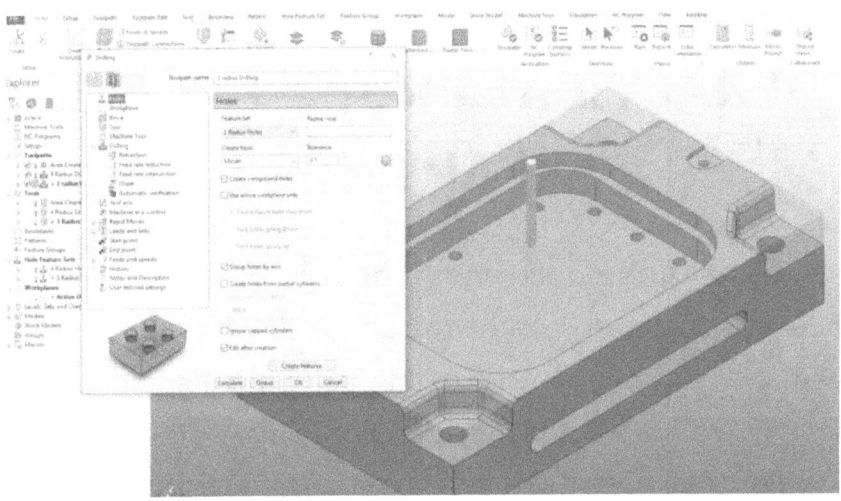

Additionally, this exercise is currently being simulated, and I will implement the concepts I have developed in the 3D area to begin the simulation and play the game. Therefore, the evacuation chamber is yours. Right now, you will observe that this section is flawless. Therefore, we will implement each operation in the simulation. Therefore, I will operate all tools except for an unusual one, as we will observe the two in the simulation. In the simulation, you will be able to see me. I will adjust the pace to a high level and supervise the operation. However, the athletes initially simulate and execute these operations in a manner that is exceedingly subtle. I will now right-click on the radius to simulate the simulation from the beginning and play the last one through various cooling options. Therefore, you will observe that our organization is. Therefore, I will simply click on it if you would like.

# PRACTICAL PROJECT CORE PART-3

We will conclude our conversation regarding the political endeavor. Simmons conducted an excellent lecture in the past, and I have inquired about him in both instances. So, in the Astro practical project, and drawing excellent back in the open, you will see that the work is already here. Now, the coding is being transformed, and the work is at a high level. Therefore, we will invert the plain and perform this task in accordance with these work planes. This will enable you to calculate the seek and approve the block. I will now defer the decision regarding the leader of the duel. Consequently, our initial operation will involve the creation of an ignorance stone, which will be a diamond that can be fired at any lecture, whether it is the same or a previous one, within a point-blank range. These sixty closing. Okay, now you will observe the totality of the situation. This is a device. I will therefore modify this moniker. Ada certification. All right. To deploy the tool, simply click on the tool and select "tool back model area clearance." All right. Currently, you can observe the use of a filthy block as an instrument. Therefore, this faucet is now being used in contemporary areas for clearance. Well, you will also be able to view the file and set it. Therefore, it was not considered a stride Gooding in the office. The finishing is also off the wall, and I will simply verify the domestic status and select "calculate." Therefore, you will observe it here; however, I would prefer to terminate it and simulate it with a fixed commitment. Now, simply select your toolbox by right-clicking, and you will be directed to this page. I will lightly simulate the ADR certification process from the beginning to the end, and you will be able to observe the results. The film is not seamless, and the incision is extremely dcep. All right. Currently, I possess an individual who is human.

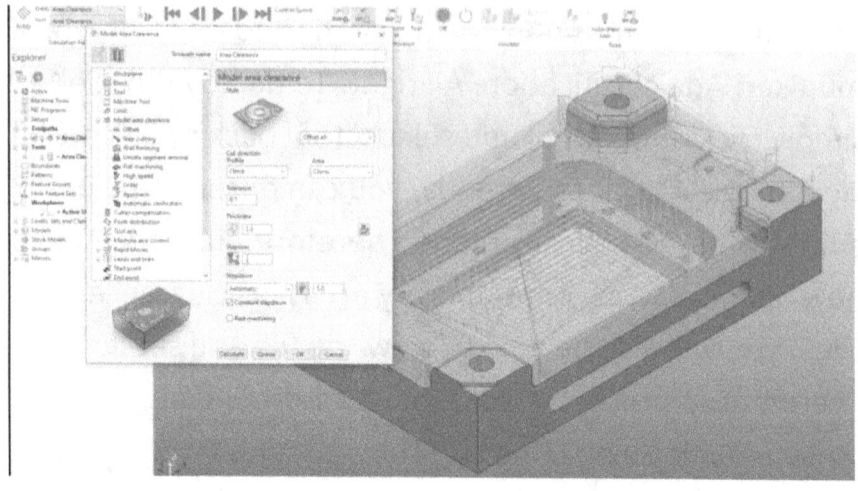

Yes, and in these cyclic settings and editable Gene these mark or one point one and step down is also going to one if this deputies file and step down this point one in calculated shape the two but ready. Therefore, I will terminate the simulation by clicking on it and then simulate from the beginning and play. This will allow you to observe the operation in action. They will transform us into this ideal shape or form. Therefore, I will depart, which will result in any US number that you may encounter in the simulation. It is flawless. Therefore, it will not be necessary to smooth over Philip. I will not conceal this instrument; however, these drilling tools and these two remain. Therefore, this radius is three and it is full. Initially, I will generate a tool with a diameter of 8 and a radius of 3. The angle is 45 degrees. All right. Also, ensure that it is closed. We will now generate the entire feature set. Therefore, use the right-click menu to generate the entire feature set. Therefore, I will also provide the name of the forward radius openings. All right. Presently, the act of right-clicking and creating is a delightful experience. The value at the bottom is minus 60. Choose your fees and submit your application. All right. Close it once more. Therefore, this is what you will observe. The instrument will be developed by me.

# PRACTICAL PROJECT CORE PART-4

The practical endeavor known as "pact" will be continued. Initially, we will establish our duel and retain our entire feature, as was discussed in a previous lecture. Additionally, in this lecture, I will select the tool. However, as I mentioned in the previous lecture, I will determine whether the field is smoothed by selecting the area clearance and the commonplace erased area clearance. You will observe this location, and we will select the Race Profile. Therefore, the entire plate is seamless; however, this operation is initially predicated on the unedited. However, it is necessary to utilize the instrument before obtaining ADF clearance. Therefore, in order to ease or complete your field, you will implement this profile in accordance with Ada clearance. Therefore, we will implement the green agreement in these instances. All right. The instrument for radius drilling now includes two openings. In drilling, the left dimension is 60. Therefore, I will simply select "calculate" and you will observe that all parties are prepared. It is so near. We will now engage in lecture check-in and check-out in the simulation.

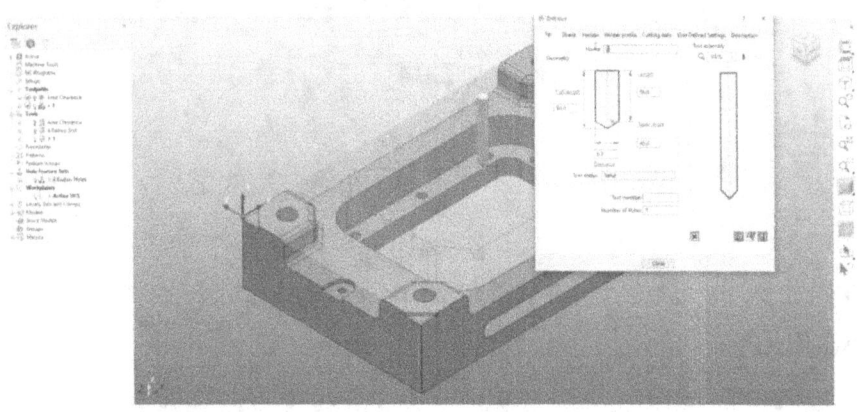

That is why I will develop a new tool for the entire three-area drill. To do so, select "drill" and assign a name to the radius drill. Currently, the amateur is six years old, and another individual holds the same value. Okay, so the entire feature suite. Immediately right-click and generate an entire feature set. Assign three locations the same name. However, this one is acceptable; it has gaps.

Therefore, create two openings by right-clicking on three significant cavities. The value is 60 less than the objectives that have been observed. Choose your fees. Now, select another visage and apply it to our poultry. You will observe that it is satisfactory. Therefore, it is imperative that you seal it completely. The feature is activated; therefore, click on the tool. However, drilling drilling is required to purchase this election. Now, the three concepts that are held in the drill are as follows: the depth is established for all three radius drills. Therefore, I will select "automatic verification" and perform the calculation to verify that all parties have submitted their applications. Therefore, I will simply close it. You will now observe that there is an additional one. Therefore, I will apply the same method to all tools. However, if these tools are concealed in the simulation, they will not be obsoleted. Okay, my pals. Ada clearance simulates the chrome stack in simulation, and the operation is conducted at maximum speed. This will be visible to you, companions, as the initial operation is executed.

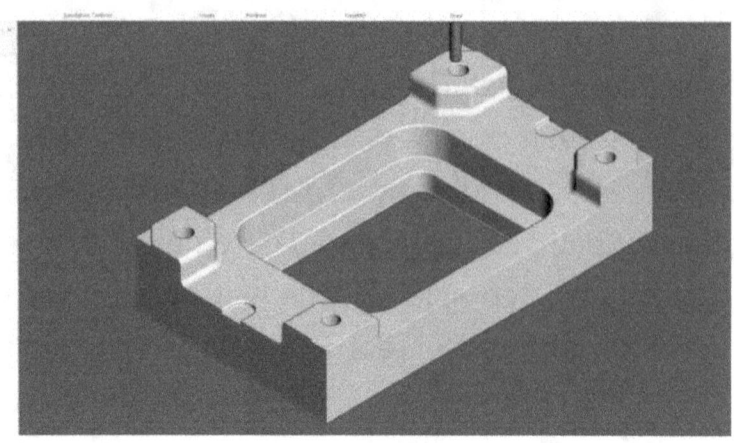

No, this one is already finished. Therefore, the radius drilling is referred to as "routine" and the least common appellation is "radius drilling." All right. No, as it pertains to drilling, simply right-click and simulate the pace and plate survey from the start of the chain. The operation is executed at an average speed of OK. This is the final individual to abandon areas that require a right-click simulation from the beginning. Generate the pace and plate to ensure that you can observe the entire operation and your object. This is the ideal solution. I will select the extant view mill and employ it. Therefore, my companions, I will conceal all of the instruments and retrieve the evidence. Therefore, I will conceal the identity of the individual who will be featured.

www.ingramcontent.com/pod-product-compliance
Lightning Source LLC
Chambersburg PA
CBHW071018240526
45469CB00006BD/1977